RIVER AND CANAL ENGINEERING

THE CHARACTERISTICS OF OPEN FLOWING STREAMS, AND THE PRINCIPLES AND METHODS TO BE FOLLOWED IN DEALING WITH THEM

BY

E. S. BELLASIS, M.Inst.C.E.

RECENTLY SUPERINTENDING ENGINEER IN THE IRRIGATION BRANCH OF THE PUBLIC WORKS DEPARTMENT OF INDIA

72 ILLUSTRATIONS

London
E. & F. N. SPON, Ltd., 57 HAYMARKET, S.W.

New York
SPON & CHAMBERLAIN, 123 LIBERTY STREET

1913

TABLE OF CONTENTS

CHAPTER I

INTRODUCTION

CHAPTER II

RAINFALL

CHAPTER III

COLLECTION OF INFORMATION CONCERNING STREAMS

CHAPTER IV

THE SILTING AND SCOURING ACTION OF STREAMS

CHAPTER V

METHODS OF INCREASING OR REDUCING SILTING OR SCOUR

CHAPTER VI

WORKS FOR THE PROTECTION OF BANKS

CHAPTER VII

DIVERSIONS AND CLOSURES OF STREAMS

CHAPTER VIII

THE TRAINING AND CANALISATION OF RIVERS

CHAPTER IX

CANALS AND CONDUITS

CHAPTER X

WEIRS AND SLUICES

CHAPTER XI

BRIDGES AND SYPHONS

CHAPTER XII

DRAINAGE AND FLOODS

CHAPTER XIII

RESERVOIRS AND DAMS

CHAPTER XIV

TIDAL WATERS AND WORKS

CHAPTER XV

RIVER BARS

PREFACE

The object of this book is to describe the principles and practice adopted in the Engineering of Open Streams. If the book seems to be somewhat small for its object, it will, it is hoped, be found that this is due to care in the arrangement and wording.

Sources of information have been acknowledged in the text, but special mention may be made of lectures given by Professor Unwin at Coopers Hill College, of Harcourt's large work on *Rivers and Canals*, of the papers[1] by Binnie on rainfall, by Shaw on the closing of the river Tista, by Harcourt on movable weirs and on estuaries, by Strange on reservoirs, and by Ottley and Brightmore, Gore and Wilson, and Hill on the stresses in masonry dams; of the articles by Bligh[2] on weirs with porous foundations and by Deacon[3] on reservoir capacity, of the Indian Government paper by Spring on "River Control on the Guide Bank System," and of the Punjab Government paper containing Kennedy's remarks on silting and scour in the Sirhind Canal. The two papers last mentioned are not easily accessible, and they contain matter of great interest. The important points, often obscured by masses of detail or figures, have been extracted.[4]

[1] *Min. Proc. Inst. C.E.* [2] *Engineering News.*

[3] *Encyclopædia Britannica.*

[4] The paper by Spring—in size it is a book—will repay perusal by engineers engaged on railway bridges over large shifting rivers. London Agents, Constable & Co.

Silting and scour (CHAP. IV.) had already been dealt with in *Hydraulics*,[1] but some further information has since come to light and the subject has been treated afresh and the matter re-written.

[1] *Hydraulics with Working Tables*. Spon, 1912.

E. S. B.

CHELTENHAM, 1st *May* 1913.

RIVER AND CANAL ENGINEERING

CHAPTER I

INTRODUCTION

1. **Preliminary Remarks.**—River and Canal Engineering is that branch of engineering science which deals with the characteristics of streams flowing in open channels, and with the principles and methods which should be followed in dealing with, altering, and controlling them. It is not necessary to make a general distinction between natural and artificial streams; some irrigation canals or other artificial channels are as large as rivers and have many of the same characteristics. Any special remarks applicable to either class will be given as occasion requires.

2. **Résumé of the Subject.**—Chap. II. of this book deals with the collection of information concerning streams, a procedure which is necessary before any considerable work in connection with a stream can be undertaken, and often before it can even be decided whether or not it is to be undertaken. Chap. III. deals with rainfall, and describes how rainfall figures and statistics can be utilised by the engineer in dealing with streams.

Chap. IV. explains the laws of silting and scouring action, a subject of great importance and one to which the attention ordinarily given is insufficient. The general characteristics of streams, being due entirely to silting or scouring tendencies, are included in this chapter. CHAP. V. describes how silting or scouring may be, under some circumstances, artificially induced or retarded.

CHAP. VI. deals with various methods of protecting banks against erosion or damage. CHAP. VII. treats of diversions or the opening out of new channels, and with the opposite of this, viz. the closing of channels, a feat which, when the stream is flowing, is sometimes very difficult to achieve. This chapter also deals with dredging and excavation.

CHAP. VIII. discusses the subject of the training of streams, a class of work which is generally undertaken in order to make them navigable or to improve their existing capacities for navigation, but may be undertaken for other reasons. The main features of this kind of work are the narrowing and deepening of the stream, often the reduction of the velocity and slope, and generally the raising of the water-level. In this kind of work a channel may be completely remodelled and even new reaches constructed. CHAP. IX. deals with artificial channels of earth or masonry, and includes navigation canals.[1]

In CHAPS. X. and XI. the chief masonry works or isolated structures—as distinguished from general works which extend over considerable lengths of channel—are dealt with, and those principles of design discussed which affect the works in their hydraulic capacities. General principles of design applicable to all kinds of works, such as the thicknesses of arches or retaining

[1] Irrigation canals are dealt with in *Irrigation Works* (Spon, 1913).

walls, are not considered; they can be found in books on general engineering design.

Chap. XII. treats of storm waters and river floods, and shows how works can be designed for getting rid of flood water and how floods can be mitigated or prevented, one of the chief measures, the widening of the channel and the lowering of the water-level, being the opposite of that adopted for training works. Embankments for stopping flooding are also dealt with. Chap. XIII. deals with reservoirs, including the design of earthen and masonry dams.

Chaps. XIV. and XV. deal with tidal waters, river mouths and estuaries, and works in connection with them, viz. the training of estuaries and the methods of dealing with bars, the object being in all cases the improvement of the navigable capacities of the channels.

3. **Design and Execution of Works.**—After obtaining full information concerning the stream to be dealt with, careful calculations are, in the case of any large and important work, made as to the effects which will be produced by it. These effects cannot always be exactly foreseen. Sometimes matters can be arranged so that the work can be stopped short at some stage without destroying the utility of the portion done, or so that the completed work can be altered to some extent.

In works for controlling streams there is, as will appear in due course, a considerable choice of types of work and methods of construction. In practice it will generally be found that there are, in any particular locality, reasons for giving preference to one particular type or kind of work or, at all events, that the choice is limited to a few of them, either because certain kinds of materials and appliances can be obtained more cheaply and readily than others, or because works of a particular type have already been successfully adopted there, or because the people of the district are accustomed to certain classes of work or methods of construction. In out-of-the-way places it is often undesirable to avoid any type of work which cannot be quickly repaired or readily kept in order by such means as exist near the spot.

It is sometimes said that perishable materials, such as trees, stakes, and brushwood, cannot produce permanent results. They can produce results which will last for a long time and which may even be permanent. By the time the materials have decayed, the changes wrought may have been very great, deposits of shingle or silt may have occurred and become covered with vegetation, and there may be little tendency for matters to revert to their former condition. If the expense of using more lasting materials had had to be incurred, the works might never have been carried out at all. On the Mississippi enormous quantities of work have been done with fascines.

4. **The Hydraulics of Open Streams.**—When any reach of a stream is altered, say by widening, narrowing, or deepening, so that the water-level is changed, there will also be a change in the water-level, a gradually diminishing change, for some distance upstream of the reach. Also in the lowest portion of the reach the change will gradually diminish and it will vanish at the extreme downstream end of the reach. In the next lowest reach there is no change. Thus if it is desired that the change in the water-level shall take full effect throughout the whole of a reach, the change in the channel must be carried further down. If a weir is

built there is no change in the water-level downstream of it except such as may be due to loss of water in the reach upstream of it. The above points are mentioned here because, although they are really questions of hydraulics, they are of much importance and of very general application.

Matters connected with the hydraulics of open streams seem to lend themselves in a peculiar way to loosely expressed remarks and fallacious opinions. The set of a stream towards a bank is sometimes supposed to profoundly affect the discharge of a diversion or branch. Its effect is simply that of "velocity of approach," which, as is well known, is quite small with ordinary velocities, and is merely equivalent to a very small increase of head. Narrow bridges or other works are sometimes said to seriously "obstruct" a stream without any observations being made of the fall in the water surface through the bridge. This fall is the only measure of the real obstruction.[1]

[1] See also Appendix A.

CHAPTER II

RAINFALL

1. **Rainfall Statistics.**—The mean annual rainfall varies very greatly according to the locality. In England it varies from about 20 inches at Hunstanton in Cambridgeshire, to about 200 inches at Seathwaite in Cumberland; in India, from 2 or 3 inches in parts of Scinde, to 450 inches or more at Cherrapunji in the Eastern Himalayas.

Rain is brought by winds which blow across the sea. Hence the rainfall in any country is generally greatest in those localities where the prevailing winds blow from seaward, provided they have travelled a great distance over the sea. Rainfall is greater among hills than elsewhere, because the temperature at great elevations is lower. Currents of moist air striking the hills are deflected upwards, become cooled, and the water vapour becomes rain. This process, if the hills are not lofty, may not produce its full effect till the air currents have passed over the hills, and thus the rainfall on the leeward slopes may be greater than elsewhere, but on the inner and more lofty ranges the rainfall is generally greatest on the windward side.

Thus the rainfall may vary greatly at places not far apart. An extreme instance of this occurs in the

Bombay hills, where the mean annual falls at two stations only ten miles apart are respectively 300 inches and 50 inches.

In temperate climates the rainfall is generally distributed over all the months of the year; in the tropics the great bulk of the rain often falls in a few months.

The fall at any one place varies greatly from year to year. To obtain really reliable figures concerning any place, observations at that place should extend over a period of thirty to thirty-five years. The figures of the mean annual fall will then probably be correct to within 2 per cent. The degree of accuracy to be expected in results deduced from observations extending over shorter periods is as follows:—

No. of years .	25	20	15	10	5
Error per cent. .	3	3¼	5	8	15

These figures were deduced by Binnie (*Min. Proc. Inst. C.E.*, vol. cix.) from an examination of rainfall figures obtained over long periods of time at many places scattered over the world. The errors may, of course, be plus or minus. They are the averages of the errors actually found, and are themselves subject to fluctuations. Thus the 15 per cent. error for a five-year period may be 16 or 13, the 8 per cent. error for a ten-year period may be 8½ or 7½, with similar but minute fluctations for the other periods.

Binnie's figures also show that the ratio of the fall at any place in the driest year to the mean annual fall, averages ·51 to ·68, with a general average of ·60, and that the ratio in the wettest year to the mean annual fall averages 1·41 to 1·75, with a general average of 1·51. For India the general averages are ·50 and 1·75.

These figures are useful as a means of estimating the probable greatest and least annual fall, but they are averages for groups of places. The greatest fall at any particular place may occasionally be twice the mean annual fall. At some places in India, in Mauritius, and at Marseilles it has been two and a half times the mean annual fall. The least annual fall may, in India, be as low as ·27 of the mean. In England the fall in a dry year has, once at least, been found to be only ·30 of the mean annual fall. The mean fall (average for all places) in the three driest years is, from Binnie's figures, about ·76 of the mean annual fall. The figures given above, except when a particular country is mentioned, apply to all countries and to places where the rainfall is heavy, as well as to those where it is light. But in extremely dry places the fluctuations are likely to be much greater. At Kurrachee, with a mean annual fall of only 7·5 inches, the fall in a very wet year has been found to be 3·73 times, and in a very dry year only ·07 times the mean annual fall.

In the United Kingdom the probable rainfall at any place in the driest year may be taken as ·63 of the mean annual fall. For periods of two, three, four, five and six consecutive dry years, the figures are ·72, ·77, ·80, ·82, and ·835. These figures are of importance in calculations for the capacity of reservoirs (CHAP. XIII., *Art.* 2).

When accurate statistics of rainfall are required for any work, the rainfall of the tract concerned must be specially studied and local figures obtained for as many years as possible. Very frequently it is necessary to set up a rain-gauge, or several if the tract is extensive or consists of several areas at different elevations.

Sometimes there is only a year or so in which to collect figures. In this case the ratio of the observed fall to that, for the same period, at the nearest station where regular records are kept, is calculated. This ratio is assumed to hold good throughout, and thus the probable rainfall figures for the new station can be obtained for the whole period over which the records have been kept at the regular station. The volumes of the British Rainfall Organisation contain a vast amount of information regarding rainfall. For a large area there should be one rain-gauge for every 500 acres, for a small area more. In the case of a valley there should be at least three gauges along the line of the deepest part—one at the highest point, one at the lowest, and one midway as regards height—and two gauges half way up the sides and opposite the middle gauge (*Ency. Brit.*, Tenth Edition, vol. xxxiii.). Some extra gauges may be set up for short periods in order to see whether the regular gauges give fair indications of the rainfall of the tract. If they do not do so some allowances can be made for this.

2. **Available Rainfall.**—The area drained by a stream is called its "catchment area" or "basin." The available rainfall in a catchment area is the total fall less the quantity which is evaporated or absorbed by vegetation. The evaporation does not chiefly take place directly from the surface. Rain sinks a short distance into the ground, and is subsequently evaporated. The available rainfall does not all flow directly into the streams. Some sinks deep into the ground and forms springs, and these many months later augment the flow of the stream and maintain it in dry seasons. The available rainfall of a given catchment area is known as the "yield" of that area.

Estimation of the available rainfall is necessary chiefly in cases where water is to be stored in reservoirs for town supply or irrigation. The ratio of the available to the total rainfall depends chiefly on the nature and steepness of the surface of the catchment area, on the temperature and dryness of the air, and on the amount and distribution of the rainfall. The ratio is far greater when the falls are heavy than when they are light. Again, when the ground is fairly dry and the temperature high—as in summer in England—nearly the whole of the rainfall may evaporate; but when the ground is soaked and the temperature low—as in late autumn and winter in England—the bulk of the rainfall runs off. In the eighteen years from 1893 to 1900 the average discharge of the Thames at Teddington, after allowing for abstractions by water companies, was in July, August, and September 12 per cent. of the rainfall—6·9 inches—in its basin, and in January, February, and March 60 per cent. of the fall which was 5·9 inches. The total fall in the year was 26·4 inches. Some rivers in Spain discharge, in years of heavy rainfall, 39 per cent., and in years of light rainfall 9 per cent. of the rainfall (*Min. Proc. Inst. C.E.*, vol. clxvii.). The discharge of a river is not always greatest in the month, or even the year, of greatest rainfall.

The table opposite gives some figures obtained by comparison of rainfall figures and stream discharges. The case of the area of 2208 acres near Cape Town is described in a paper by Bartlett (*Min. Proc. Inst. C.E.*, vol. clxxxviii.), and it is shown by figures that part of the rainfall in the rainy season went to increase the underground supply which afterwards maintained the flow in the dry season.

Place.	Catchment Area.	Period over which Observations Extended.	Total Rainfall Observed.	Available Rainfall.	Remarks.
	Acres.		Inches.	Ratio to Total.	
Nagpur, Central India .	4,224	June to September (Monsoon period).	44	·40	
„ „ .	„	„	30	·27	
Mercara, South India .	48	Whole year.	119	·37	Gravelly soil overlying granite.
King William's Town, Cape Colony . .	67,200	„	27	·21	Hills with forest and bush.
Near Cape Town, Cape Colony . . .	110	May to October (rainy season).	31·5	·51	Bare hills.
Near Cape Town, Cape Colony . . .	2,208	„	43	·40	„
Melbury Moor, Devonshire . . .	...	Whole year.	50·7	·54	
Newport, Monmouthshire	...	„	40	·40	
Newport, Isle of Wight	...	„	32	·40	
Basin of Nepean River, N.S.W. . . .	284 sq. miles	„	44·3	·44	Bare, broken ground.
Basin of Cataract River, N.S.W. . . .	70 sq. miles	„	54	·45	„ „

The following statement shows how the available rainfall may vary from year to year. The figures are those of a catchment area of 50 square miles on the Cataract River, New South Wales (*Min. Proc. Inst. C.E.*, vol. clxxxi.):—

Year.	Rainfall.	Available Rainfall.	Remarks.
	Inches.	Ratio to Total.	
1895	34·1	·84	Heavy rain falling on saturated area.
1896	33·7	·28	Evenly distributed fall.
1897	44·7	·49	Heavy rains in May.
1898	56·4	·45	„ „ February (15 ins.).
1899	54·9	·43	„ „ August (11·5 ins.).
1900	26·1	·50	„ „ May and July.
1901	37·4	·11	Evenly distributed fall.
1902	29·9	·06	
1903	41·7	·23	No heavy fall.

The manner in which the available rainfall may vary from month to month is shown in the following statement, which gives the figures for 1905 for the Sudbury River in Massachusetts:—

Month.	Rainfall.	Available Rainfall.
	Inches.	Percentage of fall.
January	5·3	48
February	2·2	24
March	3·2	142
April	2·7	104
May	1·3	40
June	5·0	16
July	5·5	6
August	2·7	8
September . . .	6·9	31
October	1·5	18
November . . .	2·1	23
December . . .	4·0	40
Total	42·3	Average 39·5

Rankine gives the ratio of the available rainfall to the whole fall as 1·0 on steep rocks, ·8 to ·6 on moorland and hilly pasture, ·5 to ·4 on flat, cultivated country, and nil on chalk. These figures are only rough. The figures for rocks and pastures are too high. The loss from evaporation and absorption is not proportional to the rainfall. It is far more correct to consider the loss as a fairly constant quantity in any given locality but increasing somewhat when the rainfall is great. The available rainfall in Great Britain has generally been overestimated. Sometimes it has been taken as being ·60 of the whole fall. More commonly the loss is taken to be 13 to 15 inches. This is correct for the western mountain districts, where the rainfall is about 80 inches and the soil consists chiefly of rocks partly covered with moorland or pasture. In other parts of the country, especially where flat, the loss is often 17 to 20 inches. All the above figures are, however, general averages. The proper estimation of the available rainfall at any place in any country depends a great deal on experience and judgment, and on the extent to which figures for actual cases of similar character are available. Regarding the "run-off" from saturated land during short periods, see CHAP. XII., *Arts. 1* and *2*.

3. **Measurement of Rainfall.**—A rain-gauge should be in open ground and not sheltered by objects of any kind. The ordinary rain-gauge is a short cylinder. This is often connected by a tapering piece to a longer cylinder of smaller diameter. In this the rain is stored safely and is measured by a graduated rod. The measurement can be made more accurately than if the diameter was throughout the same as at the top. In

other cases the water is poured out of the cylinder into a measuring vessel. If the rain-gauge was sunk so that the top was level with the ground, rain falling outside the gauge would splash into it and vitiate the readings unless the gauge was surrounded by a trench. Ordinarily the top of the gauge is from 1 to 3 feet above the ground. When it is 1 metre above the ground the rain registered is said to be on the average about 6 per cent. less than it should be, owing to the fact that wind causes eddies and currents and carries away drops which should have fallen into the gauge. The velocity of the wind increases with the height above the ground, and so does the error of the rain-gauge. Devices for getting rid of the eddies have been invented by Boernstein and Nipher (*Ency. Brit.*, Tenth Edition, vol. xviii.), but they have not yet come into general use. The Boernstein device is being used experimentally at Eskdalemuir. It would appear that much splashing cannot take place when the ground is covered with grass, and that in such a case the top of the gauge could be 1 foot above the ground, thus making the error very small.

If the ground is at first level, then rises and then again becomes level, a rain-gauge at the foot of the slope will, with the prevailing wind blowing up the slope, register too much, and a rain-gauge just beyond the top of the slope will register too little (*Ency. Brit.*, Tenth Edition, vol. xxxiii.).

4. **Influence of Forests and Vegetation.**—When the ground is covered with vegetation, and especially forests, the *humus* or mould formed from leaves, etc., absorbs and retains moisture. It acts like a reservoir, so that the run-off takes place slowly and the denudation

and erosion of the soil is checked. The roots of the trees or other vegetation also bind the soil together. Vegetation and forests thus mitigate the severity of floods and reduce the quantity of silt brought into the streams. They also shield the ground from the direct rays of the sun and so reduce evaporation, and thus, on the whole, augment the available rainfall. Forests render the climate more equable and tend to reduce the temperature, and they thus, at least on hills, increase the actual rainfall to some extent.

If a forest is felled and replaced by cultivation, the ploughing of the soil acts in the same way as the *humus* of the forest, and the crops replace the trees; and it has been stated that in the U.S.A. the cultivation is as beneficial as the forests in mitigating floods and checking denudation of the soil (*Proc. Am. Soc. C.E.*, vol. xxxiv.). But when forests are felled they are not, at least in hilly country, always replaced by cultivation. Measures to put a stop to the destruction of forests or to afforest or reforest bare land may enter into questions of the régime of streams or the supply of water. On the Rhine, increase in the severity of floods was distinctly traced to deforestation of the drainage area.

It is usually said that forests act as reservoirs by preventing snows from melting. This is disputed in the paper above quoted, and it is stated that in the absence of forests the snow forms drifts of enormous depth, and these melt very gradually and act as reservoirs after the snow in the forests has disappeared.

5. **Heavy Falls in Short Periods.**—When rain water, instead of being stored or utilised, has to be got rid of, it is of primary importance to estimate roughly—exact estimates are impossible—the greatest probable fall

in a short time. This bears a rough ratio to the mean annual fall. The maximum observed falls in twenty-four hours range, in the United Kingdom, generally from ·05 to ·10 of the mean annual fall—but on one occasion the figure has been ·20,—and in the tropics from ·10 to ·25. Actual figures for particular places can be extracted from the rain registers, but the probability of their being exceeded must be taken into account. The greatest fall observed in twenty-four hours in the United Kingdom is 7 inches, and in India 30 inches in the Eastern Himalayas.

But much shorter periods than twenty-four hours have to be dealt with. The following figures are given by Chamier (*Min. Proc. Inst. C.E.*, vol. cxxxiv.) as applicable to New South Wales, and he considers that they are fair guides, erring on the side of safety, for other countries:—

Duration of fall in hours .	1	4	12	24
Ratio of fall to maximum daily fall	$\frac{1}{4}$	$\frac{1}{2}$	$\frac{3}{4}$	1

The above figures are probably safe for England. For India the case is far otherwise. The following falls have been observed there:—

Period.	Fall.	Rate per Hour.	Remarks.
	Inches.	Inches.	
7 hours . .	10	1·43	
4·5 hours . .	7·7	1·7	
2 hours . .	8	4	
1 hour . .	5	5	
20 minutes .	1·6	4·8	
10 minutes .	1	6	

The falls of 1 inch in ten minutes were frequently observed near the head of the Upper Jhelum Canal, a place where the annual raiufall is not more than 30 inches (see also CHAP. XII., *Art. 1*). In some parts of the Eastern Himalayas, where 30 inches of rain has fallen in a day, it is possible that 8 inches may have fallen in an hour. In England 4 inches has fallen in an hour. The heaviest falls in short periods do not usually occur in the wettest years, and they may occur in very dry years. Nor do they always occur on a very wet day.

CHAPTER III

COLLECTION OF INFORMATION CONCERNING STREAMS

1. **Preliminary Remarks.**—The information which is required concerning streams depends on the character of the stream and on the nature of the work which is to be done. For the present let it be supposed that the stream is large and perennial. Other kinds of streams will be dealt with in *Arts. 6* and *7*. In dealing with a large perennial stream it is nearly always necessary to know the approximate highest and lowest water-levels, and these can generally be ascertained by local inquiry, combined with observations of water marks; but a higher level than the highest known and a lower level than the lowest known are always liable to occur, and must to some extent be allowed for. If navigation exists or is to be arranged for, the highest and lowest levels consistent with navigation must be ascertained. The highest such level depends chiefly on the heights of bridges. A plan to a fairly large scale is also necessary in most cases.

If an embankment to keep out floods is to be made along a river which is so large that its flood-level cannot be appreciably affected by the construction of the work, it may be necessary to obtain information only as to the actual flood-levels, and as to the extent to which the stream is liable to erode its bank. If a length of the bank of a stream has to be protected against scour, it is necessary to know of what materials the bed and bank are composed, and whether the channel is liable to changes and to what extent. It is also desirable to know to what extent the water transports solids, if any. In some kinds of protective work these solids are utilised.

But in cases where the stream is to be much interfered with, it is necessary to have full information concerning it, not only as regards water-levels, changes in the channel, and transport of solids, but as regards the longitudinal profile and cross-sections, and the discharges corresponding to different water-levels. The collection of some of this information, particularly as to the water-levels and discharges at different times of the year and in floods, may occupy a considerable time.

Methods of ascertaining the quantity of silt carried in the water of a stream are described in CHAP. IV., *Art. 4.* Remarks regarding the other kinds of information required—the stream being still supposed to be large and perennial—are given in *Arts. 2* to *5* of this chapter. The degree of accuracy required in the information depends, however, on the importance of the work, and sometimes the procedure can be simplified. Detailed remarks on gauges and on the instruments used and methods adopted for observing discharges and surface slopes, are given in *Hydraulics*, CHAP. VIII. and Appendix H.

2. **Stream Gauges.**—Unless the stream being dealt with is an artificial one, it is unlikely that the flow in the reach with which the work is concerned will be uniform. The rise and fall of the water at one place

cannot therefore be correctly inferred from those at another. It will be desirable to have two gauges, either read daily or else automatically, recording the water-level, one near each end of the reach concerned, with intermediate gauges if the reach is very long. If, in or near the reach, there is already a gauge which has been regularly read, it may be sufficient to set up only one new gauge, and to read it only for such a period of time as will give a good range of water-level, and to compare the readings with those of the old gauge. The readings of the new gauge for water-levels outside the range of those observed can then be inferred, but if the stream is very irregular this may involve some trouble (*Art. 4*).

In the case of a large stream which shifts its course, the reading of a gauge does not give a proper indication of the water-level. In other words, the distance of the gauge from the two ends of the reach is subject to alteration. The case is the same as if the stream was stable and the gauge was shifted about. In such a stream there ought, if accuracy is required, to be a group of two or more gauges for each point where there would be only one if the stream was not a shifting one. Also, owing to erosion of the bank or the formation of a sandbank, it may often be necessary to shift the gauge. When possible it should be kept in a fixed line laid down at right angles to the general direction of the stream. When shifted, its zero level should be altered in such a way that the reading at the new site at the time of shifting is the same as it was before shifting. When the gauge is moved back to the original site its zero should be placed at its original level, though this may give rise to a sudden jump in the reading for the reason given in the first sentence of this paragraph.

3. **Plan and Sections.**—Making a survey and plan, and laying down on it the lines for longitudinal and cross-sections, and taking levels for the sections, are ordinary operations of surveying. If any land is liable to be flooded, its boundaries should be shown on the plan and on some of the cross-sections. Unless the water is shallow, it is necessary to obtain the bed levels from the water-level by soundings, the level of a peg at the water-level having been obtained by levelling. All the sections should show the water-level as it was at some particular time, but the water-level will probably have altered while the survey was in progress, and allowance must be made for this. The pegs at all the cross-sections and on both banks of the stream—for the water-levels at opposite banks may not be exactly the same—may, for instance, be driven down to the water-level when it is steady, and thereafter any changes in it noted and the soundings corrected accordingly.

In order to ascertain what changes are occurring in the channel it may be necessary to repeat the soundings at intervals and, if there is much erosion of the bank, to make fresh plans.

4. **Discharge Observations.**—For a large stream it is necessary to observe the discharges by taking cross-sections and measuring the velocity. If there is a sufficient range of water-levels, it will be possible to make actual observations of a sufficient number of discharges. If soundings cannot, owing to the depth or velocity, be taken at high water, they must be inferred from those previously taken, but this does not allow for changes in the channel, which are

sometimes considerable and rapid. If there is not a sufficient range of water-level, the discharges for some water-levels must be calculated from those at other water-levels. In this case observations of the surface slope will be required, and the discharge site should be so selected that no abrupt changes in the channel will come within the length over which the observations are to extend. This length should be such that the fall in the water surface will be great enough to admit of accurate observation. If the cross-section of the stream is nearly uniform throughout the whole of this length, or if it varies in a regular manner, being greatest at one end of the length and least at the other end, the differences in the areas of the two end sections being not more than 10 or 12 per cent., then the velocity and cross-section of the stream can be observed in the usual manner at the centre of the length; but otherwise they should be observed at intervals over the whole length, or at least in two places, one where the section is small and one where it is great, and the mean taken. Or the velocity can be observed at only one cross-section and calculated for the others by simple proportion and the mean taken. The coefficient C can then be found from the formula $C = \frac{V}{\sqrt{R\ S}}$. To find the discharge for a higher or lower water-level, the change in the value of C corresponding to the change in R can be estimated by looking out the values of C in tables, and the discharge calculated by using the new values of C and R and the new sectional area, S remaining unaltered. But if the channel is such that, with the new water-level, a change in S is likely to have occurred, this change must be allowed for. Any such change will be due to the changed relative effects of irregularities, either in the length over which the observations extended or down-stream of that length. The effect of irregularities in the bed is greatest at low water. The effect of lateral narrowings is greatest at high water. Since a change of 10 per cent. in S causes a change of only 5 per cent. in V, it will usually suffice to draw on the longitudinal section the actual water surface observed and to sketch the probable surface for the new water-level. If the whole channel is fairly regular for a long distance down-stream of the discharge site, no slope observations need be made nor need several sections be taken in order to find V. The changed value of C should, however, be estimated in the manner above indicated. For this purpose any probable value of S will suffice.

5. **Discharge Curves and Tables.**—Ordinarily it will be possible, by plotting the observed discharges as ordinates, the gauge readings being the abscissæ, to draw a discharge curve and from it construct a discharge table. Unless the channel is of firm material and not liable to change, there are likely to be discrepancies among the observed discharges, so that a regular discharge curve will not pass through all the plotted points. If the discrepancies are not serious, they can be disregarded and the curve drawn so as to pass as near as possible to all the points, but otherwise trouble and uncertainty may arise. The soundings should be compared in order to see whether changes have occurred in the channel. If such changes do not account for the discrepancies, the cause must be sought for in some of the recorded velocities. If no sources of error in these can be found, such as wind, it is possible that the velocity has been affected by a change in the surface

slope owing to some change in the channel downstream of the length. Failing this explanation, the discrepancies must be set down to unknown causes. With an unstable channel and where accuracy is required, the sectional areas and velocities should be regularly tabulated or plotted so that changes may be watched and investigated. To do this it may be necessary to take surface slope observations, or to set up extra gauges which will show any changes in the slope.

If, downstream of the discharge site, there is any place where affluents come in and bring varying volumes of water, or where gates or sluices are manipulated, and if the influence of this extends up to the discharge site, the water-level there no longer depends only on the discharge, and a discharge table must be one with several columns whose headings indicate various conditions at the place where the disturbances occur.

In order to show how the gauge readings and discharges vary from day to day throughout the year, a diagram should be prepared showing the gauge readings and discharges as ordinates, the abscissæ being the times in days starting from any convenient date as zero. Such a diagram, showing only gauge readings, is given in fig. 56, Chap. XII.

6. **Small Streams.**—Small streams will now be considered, those, for instance, which are too small to be navigable and which occasionally run dry or nearly dry. If the water of the stream is to be stored for water supply, power or irrigation purposes, full information as to discharges and silt carried will be required. If the stream is small enough the discharges can be ascertained by means of a weir of planks. The discharge is then

known from the gauge readings. Cross-sections and large scale plans will not be required unless the stream is to be altered or embanked. If the water, instead of being stored, is to be got rid of, as in drainage work, the only information required as to discharges is the maximum discharge. Large scale plans, sections, or information as to silt or water-levels (except as a means of estimating the discharge) will not be required unless the stream is to be altered or embanked.

In all these cases of small streams the information required is generally, as has been seen, less than in the case of large perennial streams, but it is generally more difficult to obtain. If the stream is ill-defined or its flow intermittent, especially if it is also very small and the place sparsely inhabited, it may be difficult to obtain any discharge figures except those based on figures of rainfall. The method of obtaining such figures has been stated in CHAP. II. The figures required are those of the annual and monthly fall when the water is to be stored, and those of the greatest fall in a short period when the water is to be got rid of. Of course a plan of the catchment area is required.

7. **Intermittent Streams.**—In the case of large streams whose flow is intermittent, the information required will, as before, depend upon the circumstances. Such streams occur in many countries. The difficulty in obtaining information is often very great. To obtain figures of daily discharge a gauge must be set up in the stream and a register kept. The chief difficulty in an out-of-the-way place is likely to be the obtaining correct information as to the maximum discharge. Information, derived from reports or from supposed flood marks, as to the highest water-level, may be inaccurate,

and information based on rainfall figures may be extremely doubtful owing to the large size of the catchment area, the absence of rain gauges, and the difficulty, especially if the rain is not heavy, in estimating the available fall. All sources of information must be utilised and, whenever possible, observations should be made over a long period of time.

8. **Remarks.**—Very much remains to be done in collecting and publishing information concerning the ratio of the discharges to the rainfall. By observing a fall of rain and the discharge of a stream before and after the fall, it is possible to ascertain the figures for that occasion, but they will not hold good for all occasions. Continuous observations are required. The chief obstacle is the expense. Not only have measuring weirs and apparatus for automatically recording the water-level to be provided, but the weirs would often cause flooding of land involving payment of compensation. The most suitable places for making observations are those where reservoirs for water-works exist or are about to be made.

CHAPTER IV

THE SILTING AND SCOURING ACTION OF STREAMS

1. **Preliminary Remarks.**—When flowing water carries solid substances in suspension, they are known as "silt." Material is also moved by being rolled along the bed of the stream. The difference between silt and rolled material is one of degree and not of kind. Material of one kind may be rolled and carried alternately. The quantity of silt present in each cubic foot of water is called the "charge" of silt. Silt consists chiefly of mud and fine sand; rolled material of sand, gravel, shingle, and boulders. When a stream erodes its channel, it is said to "scour." When it deposits material in its channel, it is said to "silt." Both terms are used irrespective of whether the material is silt or rolled material. A stream of given velocity and depth can carry only a certain charge of silt. When it is carrying this it is said to be "fully charged."

If a stream has power to scour any particular material from its channel, it has power to transport it; but the converse is not true. If the material is hard or coherent, the stream may have far more difficulty in eroding it than in merely keeping it moving. And there is generally a little more difficulty even when the material is soft.

Silting or scour may affect the bed of a channel or the sides or both. The channel may thus decrease or increase in width or—if one bank is affected more than, or in a different manner to, the other—alter its position laterally whether or not it is altering its bed level, and *vice versa.*

The cross-section of a stream is generally "shallow," *i.e.* the width of the bed is greater than the combined submerged lengths of the sides, and the action on the bed is generally greater than on the sides.

Silting and scouring are generally regular or irregular in their action according as the flow is regular or irregular, that is, according as the channel is free or not from abrupt changes and eddies. In a uniform canal fed from a river, the deposit in the head reach of the canal forms a wedge-shaped mass, the depth of the deposit decreasing with a fair approach to uniformity. Salient angles or places alongside of obstructions are most liable to scour, and deep hollows or recesses to silt. Eddies have exceptionally strong scouring power. Immediately downstream of an abrupt change scour is often severe. An abrupt change is one, whether of sectional area or direction of flow, and whether or not accompanied by a junction or bifurcation, which is so sudden as to cause eddies. The hole scoured alongside of an obstruction may extend to its upstream side, though there is generally little initial tendency to scour there. An obstruction is anything causing an abrupt decrease in any part of the cross-section of a stream, whether or not there is a decrease in the whole cross-section, *e.g.* a bridge pier or spur.

Most streams vary greatly at different times both in volume and velocity and in the quantity of material brought into them. Hence the action is not constant. A stream may silt at one season and scour at another, maintaining a steady average. When this happens to a moderate extent, or when the stream never silts or scours appreciably, it is said to be in "permanent régime," or "stable." Most streams in earthen channels are either just stable and no more, or are unstable.

Waves, whether due to wind or other agency, may cause scour, especially of the banks. Their effect on the bed becomes less as the depth of water increases, but does not cease altogether at a depth of 21 feet, as has been supposed. Salt water possesses a power of causing mud, but not sand, to deposit.

Arts. 2, 3, and *6* of this chapter refer to action on the bed of a stream. Action on the sides will be considered in *Art. 7.*

Weeds usually grow only in water which has so low a velocity that it carries no silt to speak of, but if any silt is introduced the weeds cause a deposit. The weeds also thrive on such a deposit.

2. **Rolled Material.**—If a number of bodies have similar shapes, and if D is the diameter of one of them and V the velocity of the water relatively to it, the rolling force is theoretically as $V^2 D^2$, and the resisting force or weight as D^3. If these are just balanced, D varies as V^2, or the diameters of similarly shaped bodies which can just be rolled are as V^2 and their weights as V^6. From practical observations, it seems that the diameters do not vary quite so rapidly as they would by the above law, the weights being more nearly as V^5.

Let a stream of pure water having a depth D, and with boulders on its bed, have a velocity V just sufficient to move them very slowly. Any larger boulders would not be moved. Any smaller boulders would move more quickly. Similarly, fine sand would be rolled more quickly than coarse sand. If the velocity of the stream increases, larger boulders would be moved. Streams are thus constantly sorting out the materials which they roll. If the bed is examined it will be found that large boulders exist only down to a certain point, smaller boulders, shingle, gravel, coarse sand and fine sand following in succession.

If the water, instead of being pure, is supposed to contain silt, this may affect its velocity—it is not, however, known to do so—but, given a certain velocity, it is not likely that the rolling power of the stream is much affected by its containing silt.

It is sometimes supposed that increased depth gives increased rolling power, because of the increased pressure, but this is not so. The increased pressure due to depth acts on both the upstream and downstream sides of a body. It is moved only by the pressure due to the velocity.

When sand is rolled along the bed of a stream there is usually a succession of abrupt falls in the bed. After each fall there is a long gentle upward slope till the next fall is reached. The sand is rolled up the long slope and falls over the steep one. It soon becomes buried. The positions of the falls of course keep moving downstream. The height of a fall in a large channel is perhaps 6 inches or 1 foot, and the distance between the falls 20 or 30 feet. A fall does not usually extend straight across the bed but zigzags.

It has sometimes been said that the inclination of the bed of a stream, when high, facilitates scour, the material rolling more easily down a steep inclined plane. The inclination is nearly always too small to have any appreciable direct effect. The inclination of the surface of the stream of course affects its velocity, and this is the chief factor in the case.

A sudden rise in the bed of a stream does not necessarily cause rolled materials to accumulate there, except perhaps to the extent necessary to form a gentle slope. Frequently even this slope is not formed, especially if the rolled material is only sand. The eddies stir it up and it is carried on. The above remarks apply also to weirs or other local rises in the bed.

3. **Materials carried in Suspension.**—It has long been known that the scouring and transporting power of a stream increases with its velocity. Observations made by Kennedy have shown that its power to carry silt decreases as the depth of water increases (*Min. Proc. Inst. C.E.*, vol. cxix.). The power is probably derived from the eddies which are produced at the bed. Every suspended particle tends to sink, if its specific gravity is greater than unity. It is prevented from sinking by the upward components of the eddies. If V is the velocity of the stream and D its depth, the force exerted by the eddies generated on 1 square foot of the bed is greater as the velocity is greater, and is probably as V^2 or thereabouts. But, given the charge of silt, the weight of silt in a vertical column of water whose base is 1 square foot is as D. Therefore the power of a stream to support silt is as V^2 and inversely as D. The silt charge which a stream of depth D can carry is as $V^{\frac{1}{2}}$. V is called the "critical velocity" for that depth, and is designated as V_0.

The full charge must be affected by the nature of the silt. The specific gravity of fine mud is not much greater than that of water, while that of sand is about 1·5 times as great. Moreover, the particles of sand are far larger than the particles of mud. If two streams of equal depths and velocities are fully charged, one with particles of mud and the other with particles of sand, the latter will sink more rapidly and will have to be more frequently thrown up. They will be fewer in number. From some observations referred to by Kennedy (*Punjab Irrigation Paper*, No. 9, "Silt and Scour in the Sirhind Canal," 1904), it appears that in a fully charged stream which carried $\frac{1}{3300}$ of its volume of a mixture of mud and sand of various grades, sand of a particular degree of coarseness formed only $\frac{1}{35,000}$ of the volume of the water, but that when the same stream was clear and was turned on to a bed of the coarse sand it took up $\frac{1}{15,000}$ of its volume. It would thus appear that the full charge of silt is less as its coarseness and heaviness are greater. This is in accordance with the laws mentioned above (*Art. 2*, par. 1). See also Chap. V., *Art. 2*, last paragraph.

It is probable that fine mud is carried almost equally into all parts of the stream, whereas sand is nearly always found in greater proportion near the bed and, as before remarked, some materials may be rolled and suspended alternately. The charge of mixed silt which a stream can carry is, no doubt, something between the charge which it can carry of each kind separately, but the laws of this part of the subject are not yet fully known. From the observations above referred to, Kennedy concludes that a canal with velocity V_0 will carry in suspension $\frac{1}{3300}$ to $\frac{1}{5000}$ of its volume of silt, according as it is charged with sand of all classes or only with the heavier classes.

Let a stream be carrying a full charge of any kind of silt. Then if there is any reduction in velocity, a deposit will occur—unless there is also a reduction of depth—until the charge of silt is reduced again to the full charge for the stream. The deposit generally occurs slowly, and extends over a considerable length of channel. The heavier materials are, of course, deposited first. If a stream is not fully charged, it tends to become so by scouring its channel. It is generally believed that a stream fully charged with silt cannot scour silt from its channel, or bear any introduction of further silt. This seems to be correct in the main, but the remarks made in the latter part of the preceding paragraph must be taken into consideration.

It has been stated (*Art. 2*) that a weir or a sudden rise in the bed does not necessarily cause an accumulation of rolled material. It never causes a deposit of suspended material unless it causes a heading up and reduction of velocity to below the critical velocity.

4. **Methods of Investigation.**—The quantity of silt in water is found by taking specimens of the water and evaporating it or, if the silt is present in great quantity, leaving it to settle for twelve hours—an ounce of alum can be added for every 10 cubic feet of water to accelerate settlement—drawing off the water by a syphon, and heating the deposit to dry it. The deposit is then measured or weighed. It is best to weigh it. If clay is filled into a measure, the volume depends greatly on the manner in which it is filled in. When silt deposits in large quantities in a channel, or when heavy scour occurs, the volume deposited or scoured is ascertained by taking careful sections of the channel.

Silt is best classified by observing its rate of fall through still water. A sand which falls at ·10 feet per second is, in India, called class $\overline{\cdot 1}$, and mixed sand

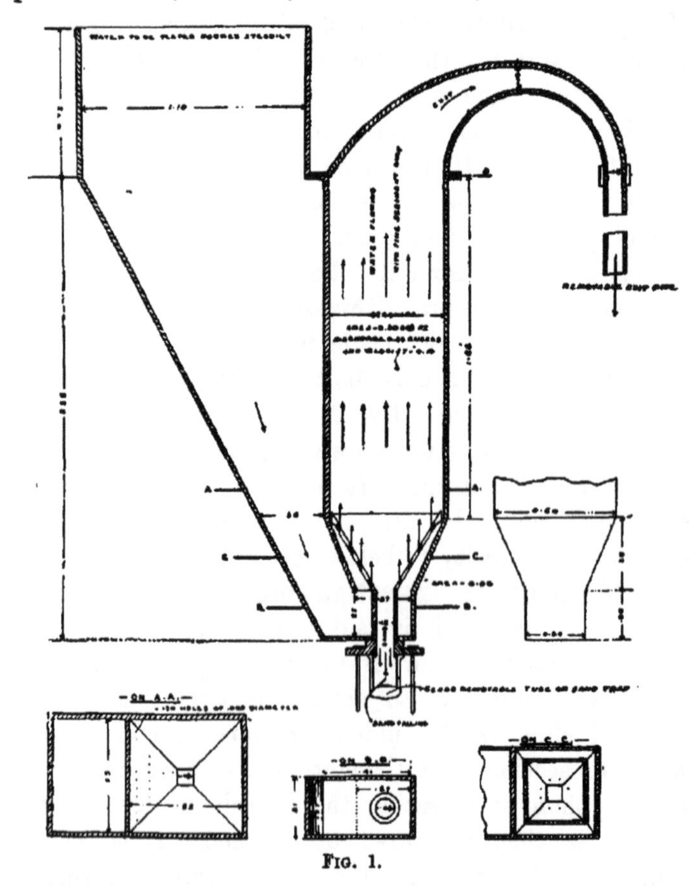

Fig. 1.

which falls at rates varying from ·1 to ·2 feet per second is called class $\frac{\cdot 1}{\cdot 2}$. Fig. 1 shows a sand separator designed by Kennedy. The scale is $\frac{1}{8}$. It has a syphon action, and the rate of flow can be altered by altering the length of the exit pipe. Suppose it is desired to measure the sand of class $\overline{\cdot 10}$ and all heavier kinds. The pipe is adjusted so as to give a velocity of ·1 foot per second to the upward flowing water, which then carries off all silt of class $\overline{\cdot 10}$ or finer. All heavier silt falls into the glass tube. It can be separated again by being mixed with water and passed through the instrument again, the velocity of flow through the instrument being increased.

The quantity of silt present at various depths can be found by pumping specimens of water through pipes. At each change of depth the pipe, delivery hose, etc., should be cleaned. Allowance must be made for the velocity of ascent of the water up the pipe. Suppose this to be 1·4 feet per second. Then the velocity of sand of class $\overline{\cdot 2}$ would be 1·2 feet per second, and the quantity of sand actually found in the water would have to be increased by one-sixth.

5. **Quantity and Distribution of Silt.** — The quantity of silt present in water varies enormously. Fine mud, even though sufficient to discolour the water, may be so small in volume that it only deposits when the water is still, and even then deposits slowly. In the river Tay, near Perth, the silt was found to be ordinarily $\frac{1}{10,000}$ of the volume of water, and at low water only $\frac{1}{28,000}$. In the river Sutlej at Rupar, near where it issues from the Himalayas, the silt in the flood season is extremely heavy. Out of 360 observations, made at various depths, during the flood seasons of four successive years, in water whose depth ranged up to 12 feet, the silt was once found to be 2·1 per cent. by weight of that of the water. It was more than 1·2 per cent. on four occasions, and more than 0·3 per cent. (or 3 in 1000) on sixty-four occasions. Generally about one-half

of the silt was clay and sand of classes finer than $\overline{\cdot 10}$, about one-third was sand of class $\frac{\cdot 1}{\cdot 2}$, and the residue was sand of class $\frac{\cdot 2}{\cdot 3}$. The sand of the river Chenab is generally coarser than that of the Sutlej. There are very great differences in the degree of coarseness of river sand. The sand in any river becomes finer and finer as the gradient flattens in approaching the sea. Sea sand has been found to be of class $\overline{\cdot 20}$. In the Sirhind Canal, which takes out from the Sutlej at Rupar, the maximum quantity of suspended silt observed in the four flood seasons was 0·7 per cent., on one occasion out of 270, and it exceeded 0·3 per cent. on twenty-five occasions. About 80 per cent. of the silt was clay.

In another part of the paper quoted, it is stated that the silt suspended in the canal water averaged, during the whole of one flood season, about $\frac{1}{1700}$ of the volume of the water. This would be about $\frac{1}{1200}$ by weight. The silt deposited in the bed of the canal, in a period of a few days, was sometimes as much as $\frac{1}{1000}$ of the water which had passed along, and occasionally as much as $\frac{1}{500}$. It was nearly all sand, only about 3 per cent. being clay. Silt of classes finer than $\overline{\cdot 1}$ gave no trouble, and were to be eliminated in future investigations. In a canal, as in a river, the sand on the bed becomes finer the further from the head.

Regarding the distribution of the silt at various depths, in water 5 to 17 feet deep, the quantity of silt near the bed may, when the charge is heavy and consists of mixed silt, be $1\frac{1}{4}$ to 3 times that at the surface. If the charge is fine mud, there is likely to be as much silt at the surface as near the bed, if sand, there may be none at the surface and little in the upper part of the stream.

In all cases single observations are likely to show extraordinarily discordant results; a number of observations must be made at each point and averaged.

6. **Practical Formulæ and Figures.**—A stream which carries silt generally rolls materials along its bed. The proportion between the quantities of material rolled and carried is never known, and this makes it impossible to frame an exact formula applicable to such cases, but Kennedy, from his observations on canals fully charged with the heavy silt and fine sand usually found in Indian rivers near the hills, arrived at the empirical formula for critical velocities

$$V = \cdot 84\ D^{\cdot 64}$$

The observations were made on the Bari Doab Canal and its branches, the widths of the channels varying from 8 feet to 91 feet, and the depths of water from 2·3 feet to 7·3 feet. The beds of these channels have, in the course of years, adjusted themselves by silting or scouring, so that there is a state of permanent régime, each stream carrying its full charge of silt, and the charges in all being about equal. From further observations referred to above (*Art. 3*, par. 2) it appears that this kind of silt forms about $\frac{1}{3300}$ of the volume of the water, and that on the Sirhind Canal, sand coarser than the $\overline{\cdot 10}$ class, formed $\frac{1}{35,000}$ of the volume of water.

The formula gives the following critical velocities for various depths :—

$D =$	1	2	3	4	5	6	7
$V_0 =$	·84	1·30	1·70	2·04	2·35	2·64	2·92

$D =$	8	9	10
$V_0 =$	3·18	3·43	3·67.

In Indian rivers not near the hills the silt carried is not so heavy, and the critical velocities are supposed to be about three-fourths of the above. Thrupp (*Min. Proc. Inst. C.E.*, vol. clxxi.) gives the following ranges of velocities as those which will enable streams to carry different kinds of silt. It does not appear that the streams would be fully charged except at the higher figure given for each case.

D = 1·0	10·0
V = 1·5 to 2·3	3·5 to 4·5 (Coarse sand).
V = ·95 to 1·5	2·3 to 3·5 (Heavy silt and fine sand).
V = ·45 to ·95	1·2 to 2·3 (Fine silt).

It cannot be said that the exact relations between D and V are yet known, but it is of great practical importance to know that V must vary with D. The precise manner in which it must vary does not, for moderate changes, make very much difference. In designing a channel a suitable relation of depth to velocity can be arranged for, and one quantity or the other kept in the ascendant, according as scouring or silting is the evil to be guarded against.

The old idea was that an increase in V, even if accompanied by an increase in D, *e.g.* simply running a higher supply in a given channel, gave increased silt-transporting power. In a stream of very shallow section this is probably correct, for V increases faster than $D^{\cdot 64}$ (*Hydraulics*, CHAP. VI., *Art. 2*). In a stream of deep section a decrease in D gives increased silt-transporting power. If the discharge is fixed, a change in the depth or width must be met by a change of the opposite kind in the other quantity. In this case widening or narrowing the channel may be proper according to circumstances. In a deep section widening will decrease the depth of water, and may also increase the velocity, and it will thus give increased scouring power. In a shallow section, narrowing will increase the velocity more than it increases $D^{\cdot 64}$. In a medium section it is a matter of exact calculation to find out whether widening or narrowing will improve matters.

If the water entering a channel has a higher silt-charge than can be carried in the channel, some of it must deposit. Suppose an increased discharge to be run, and that this gives a higher silt-carrying power and a smaller rate of deposit per cubic foot of discharge, it does not follow that the deposit will be less. The quantity of silt entering the channel is now greater than before. Owing to want of knowledge regarding the proportions of silt and rolled material, and to want of exactness in the formulæ, reliable calculations regarding proportions deposited cannot be made.

The channels in which the observations above referred to were made have all assumed nearly rectangular cross-sections, the sides having become vertical by the deposit on them of finer silt; but the formula probably applies approximately to any channel if D is the mean depth from side to side, and V the mean velocity in the whole section.

If the ratio of V to D differs in different parts of a cross-section, there is a tendency towards deposit in the parts where the ratio is least, or to scour where it is greatest. There is a tendency for the silt-charge to adjust itself, that is, to become less where the above ratio is less, but the irregular movements of the stream cause a transference of water among all parts, and this tends to equalise the silt-charge.

Dubuat gives the following as the velocities close to the bed which will enable a stream to scour or roll various materials. The bed velocity is probably less than the mean velocity in the ratio of about ·6 to 1 in rough channels, and about ·7 to 1 in smooth channels :—

Gravel as large as peas	·70	feet per second
,, ,, French beans	1·0	,, ,, ,,
,, 1 inch in diameter	2·25	,, ,, ,,
Pebbles 1½ inch in diameter	3·33	,, ,, ,,
Heavy shingle	4·0	,, ,, ,,
Soft rock, brick, earthenware	4·5	,, ,, ,,
Rock of various kinds	6·0	,, ,, ,,
	and upward.	

The figures for brick, earthenware, and rock can apply only to materials of exceedingly poor quality. Masonry of good hard stone will stand 20 feet per second, and instances have occurred in which brickwork has withstood a velocity of 90 feet per second without injury so long as the water did not carry sand and merely flowed along the brickwork. If there are abrupt changes in the stream, causing eddies, or if there is impact and shock, or if sand, gravel, shingle, or boulders are liable to be carried along, velocities must be limited.

7. **Action on the Sides of a Channel.**—It has been seen that the laws of silting and scour on the bed of a channel depend on the ratio of the depth to the velocity. The same laws probably hold good in the case of a gently shelving bank, so that here again V ought to vary as $D^{\cdot 64}$. The velocity near the angle where the slope meets the water surface seems to decrease faster than $D^{\cdot 64}$. At all events, silt tends to deposit in the angle and the slope to become steep.

When the slope is steep the law seems to be different, the tendency for deposit or scour to occur on the bank depending on the actual velocity without much relation to the depth. The velocity very near to a steep bank is always low relatively to that in the rest of the stream. Thus there is often a tendency for silt to deposit on the bank, especially in the upper part, and for the side to become vertical except for a slight rounding at the lower corner. A bank may receive deposits when the bed may be receiving none, and it may have a persistent tendency to grow out towards the stream. The growth of the bank is generally regular, the line of the bank being preserved, but it may be irregular, especially if vegetation, other than small grass, becomes established on the new deposits.

When scour of the sides of a channel occurs it may occur by direct action of the stream on the sides near the water-level, or by action at or near the toe of the slope, which causes the upper part of the bank to fall in. Such falling in is generally more or less irregular, and the bank presents an uneven appearance. The fallen pieces of bank may remain, more or less intact, especially if they are held together by the roots of grasses, etc., where they fell, and prevent further scour occurring along the toe of the slope. Falling in of banks is most liable to occur in large streams and with light soils. It may be caused by the waves which are produced by steamers and boats or, especially in broad streams, by wind. The action on the banks at bends is discussed in *Art. 8*.

Thus in designing a channel according to the principles laid down in *Art. 6*, the question of action on the sides of the channel has to be dealt with as follows. Whether or not the velocity is to be low, relatively to the depth, *i.e.* whether or not deposit on the bed is more likely to occur than scour, care can be taken not to make it actually too low, and not to make it actually too high, particularly if the soil is light and friable. With ordinary soils a mean velocity of 3·3 feet per second in the channel is generally safe as regards scour of the sides. Any velocity of more than 3·5 feet per second may give trouble. A velocity of less than 1 foot per second is likely to give rise to deposit on the sides.

In channels in alluvial soils the falling in of banks is sometimes said to occur more when the stream is falling than at other times. This has been noticed on both the Mississippi and the Indus. The cause has been said to be the draining out of water which had percolated into the bank, the water in flowing out carrying some sand with it. The effect of this cannot however be great.

8. **Action at Bends.**—At a bend, owing to the action of centrifugal force and to cross-currents caused thereby, there is a deposit near the convex bank and a corresponding deepening—unless the bed is too hard to be scoured—near the concave bank. The water-level at the concave bank is slightly higher than at the convex bank. The greatest velocity instead of being in mid-stream is nearer the concave bank.

As the transverse current and transverse surface slope cannot commence or end abruptly, there is a certain length in which they vary. In this length the radius of curvature of the bend and the form of the cross-section also tend to vary. This can often be seen in plans of river bends, the curvature being less sharp towards the ends.

When once a stream has assumed a curved form, be

it ever so slight, the tendency is for the bend to increase. The greater velocity and greater depth near the concave bank react on each other, each inducing the other. The concave bank is worn away, or becoming vertical by erosion near the bed, cracks, falls in, and is washed away, a deposit of silt occurring at the convex bank, so that the width of the stream remains tolerably constant. The bend may go on increasing, and it often tends to move downstream.

In fig. 2 the deep places are shown by dotted lines. Along the straight dotted line there is no deep place.

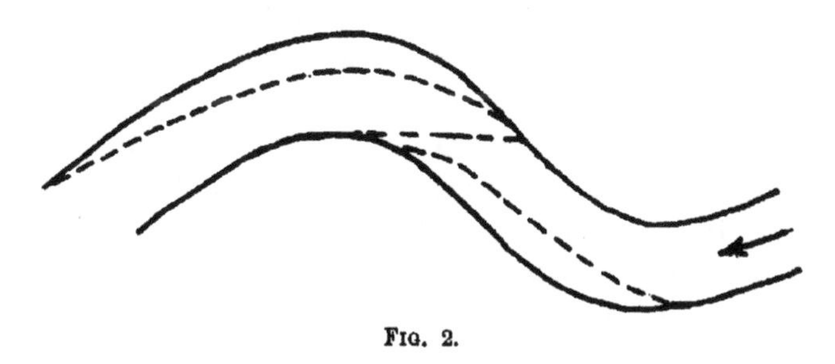

FIG. 2.

Such a line would be used for a ford. At low water it becomes a shoal. This is the chief reason why a tortuous stream at low water consists of alternate pools and rapids. It is sometimes said that deep water occurs near to a steep hard bank. Such deepening is due to bends or obstructions which give the current a set towards the bank, or it is due to irregularities in the bank which cause eddies. In a straight channel with even and regular banks there is no such deepening.

When a bend has formed in a channel previously straight, the stream at the lower end of the bend, by setting against the opposite bank, tends to cause another bend of the opposite kind to the first. Thus the tendency is for the stream to become tortuous and, while the tortuosity is slight, the length, and therefore the slope and velocity, are little affected; but the action may continue until the increase in the length of the stream materially flattens the slope, and the consequent reduction in velocity causes erosion to cease. Or the stream during a flood may find, along the chord of a bend, a direct route with, of course, a steeper slope. Scouring a channel along this route it straightens itself, and its action then commences afresh. Short cuts of this kind do not, however, occur so frequently as is sometimes believed. In some streams the bends acquire a horse-shoe shape and the neck becomes very narrow and short cuts may then occur. Otherwise they are not common. V increases only as $\sqrt{S}$, and if the country is covered with vegetation it is not easy for a stream to scour out a new channel.

The effect of bends on the velocity of a stream is not well understood. In case of a bend of 90° the increased resistance to flow when the bend is absolutely sudden (a sudden bend is known as an "elbow") amounts perhaps to $\frac{V^2}{2g}$. Whether it is greater or less in the case of a gentle bend of 90° is not known. In the case of a pipe there is a certain radius which gives a minimum resistance (*Hydraulics*, Chap. V.). The increased resistance at a bend is due partly to the fact that the maximum velocity is no longer in the centre of the stream, and partly to the fact that the velocities at the different parts of the cross-section have to be rearranged at the commencement of the bend and again at its termination. Thus the effect of a bend of 45° is a good deal more than half of that of a bend of 90°. Two bends of

45°, both in the same direction, with a straight reach between them, will cause more resistance than a single bend of 90° with the straight reach above or below the bend. If the two bends of 45° are in different directions the resistance will be still greater. A succession of sharp bends may produce a serious effect, amounting to an increase in roughness of the channel. A succession of gentle bends, of any considerable angle, cannot of course occur within a moderate length of channel.

When there is head to spare there is clearly no objection to bends, except that the bank may need protection. At a place where the bank has in any case to be protected, *e.g.* at a weir, there is no objection to an elbow.

9. **General Tendencies of Streams.** — Since the velocity is greater as the area of the cross-section is less, a stream always tends to scour where narrow or shallow, and to silt where wide or deep. The cross-section thus tends to become uniform in size. Suppose two cross-sections to be equal in size but different in shape. The velocities of the two sections will be equal. The tendency of the bed to silt will (*Art. 6*) be greater at the deeper section and, when silting has occurred on the bed, the section will be reduced and there will be a tendency to scour at the sides. Thus the cross-sections tend to become also uniform in shape. If a bank of silt has formed in a stream, the tendency is for scour to occur. There is also a tendency for silt to deposit just below the point where the bank ends. Hence a silt bank often moves downstream.

Owing to the tendency to scour alongside of, or downstream of, obstructions (*Art. 1*), it is clear that a stream constantly tends to destroy obstructions.

There is an obvious tendency for silt to deposit where the bed slope of a stream flattens, and for scour to take place where it steepens (*Hydraulics*, figs. 16 and 17, pp. 24 and 25), and thus the tendency is for the slope to become uniform.

In a natural stream flowing from hilly country to a lake or sea, the slope is steepest at the commencement and gradually flattens. There is thus a tendency for the bed to rise except at the mouth of the stream. This rising tends to increase the slope and velocity in the lower reaches, and this again enhances the tendency, described in the preceding article, of the stream to increase in tortuosity.

When a silt-bearing stream overflows its banks the depth of water on the flooded bank is probably small and its velocity very low, and a deposit of silt takes place on the bank. When the deposit has reached a certain height it acts like a weir on the water of the next flood, which flows quickly over it and, instead of raising it higher, deposits its silt further away from the stream. In this way a strip of country along the stream gradually becomes raised, the raising being greatest close to the stream. The country slopes downwards in going away from the stream. In other words, the stream runs on a ridge. If the bank becomes raised so high that flooding no longer occurs, the raising action ceases, but if, as is likely in alluvial country, the bed of the stream also rises, the action may continue and the ridge become very pronounced.

Some rivers have very wide and soft channels which are only filled from bank to bank in floods, if then. The deep stream winds about in the channel, and the rest of it is occupied by sandbanks and minor arms. The winding is the result of the velocity being too great for the channel. The streams, especially the main stream, constantly shift their courses by scouring one bank or the other. Now and then the main stream takes a short cut, either down a minor arm or across an easily eroded sandbank. This is a very different matter from a short cut across high ground. The sandbanks receive deposits of silt in floods, but are constantly being cut away at the sides. Such rivers frequently erode their banks to an extraordinary extent. The Indus sometimes cuts into its bank 100 feet or more in a day, and it may cut for half a mile or more without cessation. The tortuosity of such a stream increases as it gets nearer the sea. The actual length of the Indus in the 400 miles nearest the sea is 39 per cent. greater than its course measured along the bank. In the reach from the 600th to the 700th mile from the sea, the difference is only 3 per cent. For a detailed description of some such rivers, see *Punjab Rivers and Works*.

Sometimes general statements are made regarding silting or scour in connection, for instance, with a stream which is confined between embankments or training walls, or has overflowed its banks or is held up by a weir. It is impossible to say that any such condition, or any condition, will cause silting or scour, unless the velocity depth and silt charge are known.

CHAPTER V

METHODS OF INCREASING OR REDUCING SILTING OR SCOUR

1. **Preliminary Remarks.**—Most important works which affect the régime of a stream have some effect on its silting or scouring action, but this is not generally their chief object. Such works will be dealt with in due course, and the effects which they are likely to produce on silting or scouring will be mentioned. In the present chapter only those works and measures will be considered whose chief object is to cause a stream to alter its silting or scouring action. It does not matter, so far as this discussion is concerned, whether the object is direct, *i.e.* concerned only with the particular place where the effect is to be produced, or indirect, as, for instance, where a stream is made to scour in order that it may deposit material further down the stream. The protection of banks from scour is considered in CHAP. VI. Dredging is dealt with in CHAP. VIII.

2. **Production of Scour or Reduction of Silting.**—Sometimes the silt on the bed of a stream is artificially stirred up by simple measures, as, for instance, by scrapers or harrows attached to boats which are allowed to drift with the stream, or by means of a cylinder which has claw-like teeth projecting from its circumference and is

rolled along the bed, or by fitting up boats with shutters which are let down close to the bed and so cause a rush of water under them, or by anchoring a steamer and working its screw propeller. It is thus possible to cause a great deal of local scour, but the silt tends to deposit again quickly, and it is not easy to keep any considerable length of channel permanently scoured. The system is suitable in a case in which a local shallow or sandbank is to be got rid of and deposit of silt a little further down is not objectionable. It may be suitable in a case in which the bed is to be scoured while a deposit of silt at the sides of the channel is required, especially if some arrangement to encourage silt deposit at the sides is used (*Art. 3*, par. 4; also CHAP. VI., *Art. 3*).

Holding back the water by means, for instance, of a regulator or movable weir, and letting it in again with somewhat of a rush, will, if frequently repeated, have some effect in moving silt on in the downstream reach. Regarding the upstream reach, it has been remarked (CHAP. IV., *Art. 3*) that a weir does not necessarily cause silt deposit. If, in a stream which does not ordinarily silt, a regulator or movable weir causes, when the water is headed up, some silt deposit, the cessation of the heading up not only removes the tendency to silt, but the section of the stream, at the place where the deposit occurred, is less than elsewhere, and there is thus a tendency to scour there. If a regulator is alternately closed and opened, no permanent deposit of much consequence is likely to occur.

A stream may be made to scour its channel by opening an escape or branch. This causes a draw in the stream, and an increase in velocity for a long distance upstream of the bifurcation (*Hydraulics*, Chap. VII., *Art. 6*). This procedure is sometimes adopted on irrigation canals. The escape is generally opened in order to reduce the quantity of water passing down, but it may be opened solely to induce scour or prevent silting. The floor of the escape head is usually higher than the bed of the canal, but this does not interfere with operations except at low supplies. It may (Chap. IV., *Art. 2*) have some effect on the quantity of rolled material passed out of the escape.

If there is a weir in the river below the off-take of the canal, and if the escape runs back to the river and thus has a good fall, the scouring action in the canal may be very powerful.

If the main channel has a uniform slope throughout, the slope of its water surface is greater upstream of the escape than downstream of the escape, and there is thus an abrupt reduction of velocity and possibly a deposit of silt in the main channel below the escape. This may or may not be objectionable. In the case of an irrigation canal, it is far less objectionable than deposit in the head of the canal. The best point for the off-take of any escape or scouring channel depends on the position of the deposits in the main channel. The off-take should be downstream of the chief deposits, but as near to all of them as possible. A breach in a bank acts of course in the same way as an escape.

A stream of clear water when sent down a channel will scour it if the material is sufficiently soft. In the case of the Sirhind Canal, it has already been mentioned (Chap. IV., *Art. 3*), that when the river water became clear after the floods the proportion of coarse sand, *i.e.* sand above the $\cdot\overline{10}$ class, carried by the canal water was about $\frac{1}{15,000}$ by volume. This was in the period from 22nd September to 7th October. From 8th to 23rd October the proportion averaged $\frac{1}{32,000}$, from 24th October to 8th November $\frac{1}{44,000}$, and from 9th to 24th November $\frac{1}{85,000}$. The reason of this reduction was that the comparatively clear water kept picking up the sand from the bed and moving it on, the finer kinds being moved most quickly. As the coarse sand left on the bed became less in quantity, the water took up less. It appears, however, that the water also picked up some clay which was left, and that the total suspended silt in November was $\frac{1}{9000}$ of the water. All the observations mentioned in this paragraph appear to have been made at Garhi, 26 miles from the head of the canal.

3. **Production of Silt Deposit.**—Works or measures for causing silt deposit may be undertaken in order to cause silt deposit in specific places where it will be useful, or in order to free the water from silt. Sometimes both objects are combined.

If a stream can be turned into a large pond or low ground—a bank being built round it if necessary—it can be made to part with some or all of its silt whether rolled or suspended. Even if the pond is so large that the velocity becomes imperceptible, the whole of the suspended matter will not deposit unless it has sufficient time, but the matter which remains in the water is likely to be extremely small in amount. The silting up of marshes, pools, borrow-pits, etc., is now being effected, or should be effected, in places where mosquitoes and malaria are prevalent.

In the upper or torrential part of a stream, a high dam, provided with a sluice and a high-level waste weir, may be built across it. The space above the dam becomes more or less filled with gravel, etc. This has been done in Switzerland (*Min. Proc. Inst. C.E.*, vol. clxxi.). In the U.S.A. long weirs have been built in order to stop the progress of detritus from gold mines. Such detritus was liable to choke up rivers and damage the adjoining lands. The detritus

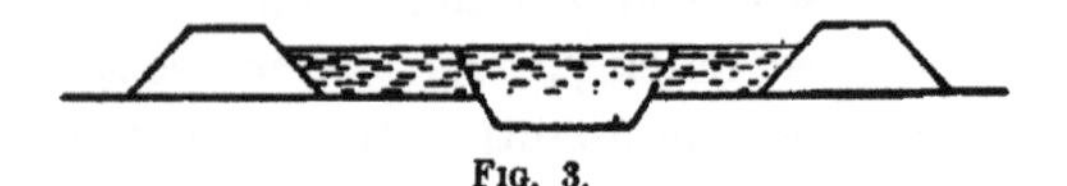

Fig. 3.

from hill torrents can also be reduced by afforestation of the hill sides.

When a stream is in embankment—irrigation channels are frequently so—the bank can be set back (fig. 3), and suspended silt will then deposit on the berms. The object of this arrangement is generally to create very strong banks in low ground. A similar plan can be adopted when the berm is only slightly below the

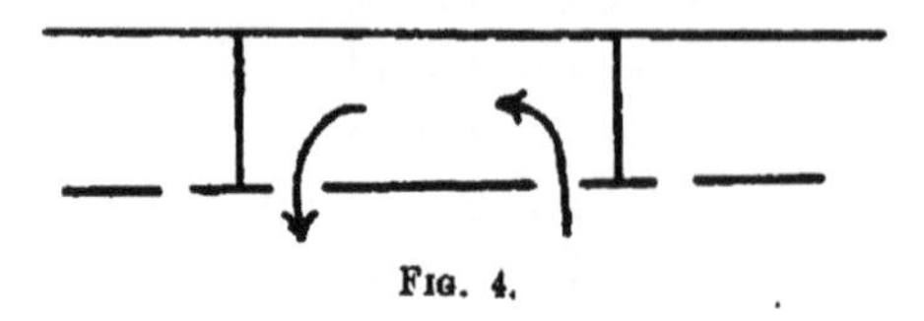

Fig. 4.

water-level and even when it is only occasionally submerged. In this case the deposit of a small bank of silt along the edge of the berm next the stream will prevent the access of fresh supplies of silt-bearing water to the parts further away. Gaps should be cut in the bank of silt at intervals, and cross banks made to form "silting tanks," as shown in fig. 4. The inlets to the tank should be large, and the outlets small, so that the water in the tank may have little velocity.

It is not, however, correct to have the outlet so small—unless the water contain very little silt—that there is very little flow through the tank. The tanks will generally be silted up most quickly by allowing a good flow through them, even though only a small proportion of the silt in the water is deposited. Regular banks arranged to form tanks on the above principle can be made behind the original banks of a canal in cases where the original banks were not, for any reason, set back.

When a channel is made in low ground and the excavation is not sufficient to make the banks, borrow-pits can be dug in the bed of the channel. Such pits should not be long and continuous, but wide bars should be left so that a number of short pits will result. These pits will trap rolled material as well as suspended silt. The object in this case is to free the water from silt and to reduce the size of the channel and thus reduce the loss of water from percolation.

On the Indus, where it has a strong tendency to shift westwards, long earthen dams or groynes are run out from the west bank across the sandbanks. One object is to cause silt deposit, and so increase the quantity of material which the river will have to cut away, but whether this result is achieved is doubtful. The sand-banks receive deposits in any case. A groyne may increase the deposit on its upstream side, but it cuts off the flood water from its downstream side and so reduces the deposit there.

4. **Arrangements at Bifurcations.**—At a bifurcation, as where a branch takes off from a canal, it is possible to reduce the quantity of rolled material entering the canal by raising its bed or constructing

a weir or "sill" in its head. This arrangement may have great effect in excluding boulders, shingle, or gravel. As regards rolled sand, it has much less effect than might be expected (CHAP. IV., *Art. 2*). If the canal is reduced in width (fig. 5) there will be eddies below the bed level of the branch. They will stir up the sand and some of it will enter the branch. If the canal is not reduced in width, eddies will be produced in the surface water, and they will affect the bed.

The above remarks apply also to the case of a canal

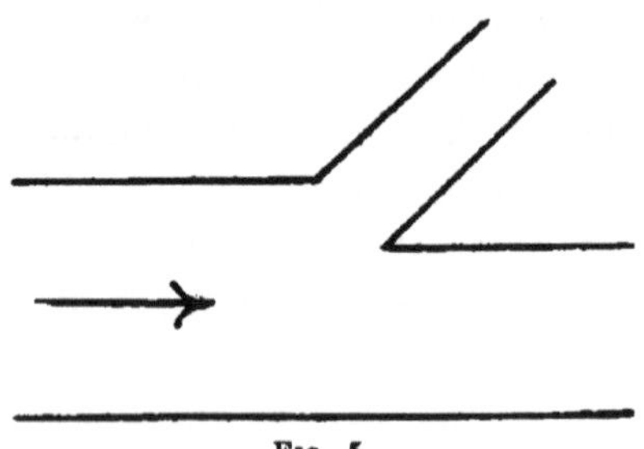

FIG. 5.

taken off from a river when there are no works in the river.

5. **A Canal with Headworks in a River.**—In the case of a canal taking off from a river and provided with complete headworks, it is possible to do a great deal more. The case of the Sirhind Canal, already referred to (CHAP. IV., *Arts. 5* and *6*), is a notable example. The canal (fig. 6) is more than 200 feet wide, the full depth of water 10 feet, and the full discharge about 7000 cubic feet per second. In 1893 when the irrigation had developed, and it became necessary to run high supplies in the summer—July, August, and part of September—the increase in the silt deposit threatened to stop the working of the canal.

In the autumn and winter, say from 25th September

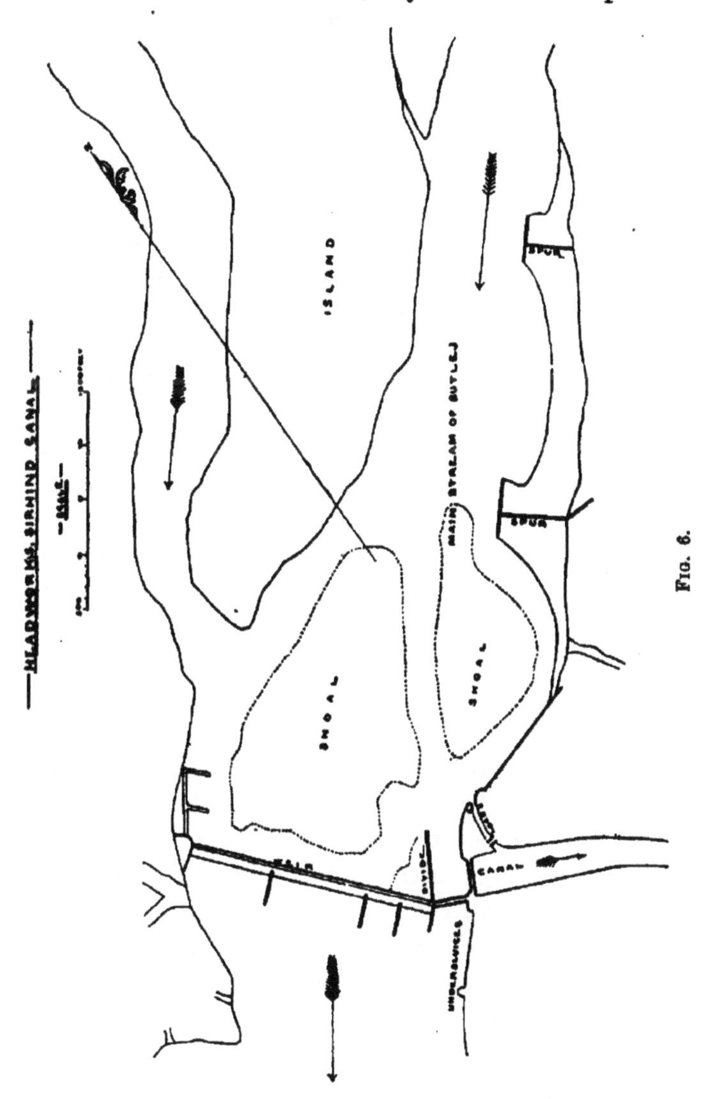

Fig. 6.

to 15th March, the water entering the canal is clear

and much of the deposit was picked up by it, but not all. In the five years 1893 to 1897 inclusive, the following remedial measures were adopted. Increased use was made of the escape at the twelfth mile. This did some good, but there was seldom water to spare. In 1893 to 1894 the sill of the regulator was raised to 7 feet above the canal bed, and it was possible to raise it 3 feet more by means of shutters. This had little effect. The coarsest class of sand was ·4, and the velocity of the water, even of that part of it which came up from the river bed and passed over the sill, was over 2 feet per second, so that all sand was carried over. In 1894 to 1895 the divide wall, which had been only 59 feet long, was lengthened to 710 feet, so as to make a pond between the divide wall and the regulator,[1] but probably the leakage through the under-sluices was often as much as the canal supply, and the water in the pond was thus kept in rapid movement and full of silt. The canal was closed in heavy floods. This did some good, but probably the canal was often closed needlessly when the water looked muddy but contained no excessive quantity of sand. The above comments on the measures taken were made by Mr Kennedy when chief engineer. The above measures did not reduce the silt deposits, but the scour in the clear water season improved, probably because higher supplies were run owing to increased irrigation. The deposit in the upper reaches of the canal, when at its maximum about the end of August of each year, was generally more than twenty million cubic feet. From the year 1900 a better system of regulation was enforced, the under-

[1] The regulator runs across the canal head; the under-sluices are a continuation of the weir, between the divide wall and the regulator.

sluices being kept closed as much as possible, so that there was much less movement in the pond and much less silt in its water. By 1904 the deposit in the canal had been reduced to three million cubic feet, and no further trouble occurred.

During the period from 20th September 1908 to 10th October 1908 the quantity of silt in the canal above Chamkour (twelfth mile) decreased from 19,325,800 cubic feet to 12,477,600 cubic feet. The quantity scoured away was 6,848,200. cubic feet. During this period no silt entered the canal. The quantity which passed out of the reach in question in suspension was 4,183,660 cubic feet, so that 2,664,540 cubic feet of material must have been rolled along the bed. The rolled material was 64 per cent. of the suspended material. During this period the Daher escape, in the twelfth mile, was open, and the mean velocity in the canal just above the escape was about 4 feet per second, the depth of water being about 10 feet. The velocity near the escape was thus greater than the critical velocity for mixed silt (CHAP. IV., *Art. 6*), and even a long way up the canal it would be in excess of the critical velocity. The water seems to have carried about $\frac{1}{1800}$ of its volume of silt. Whether the above proportions of rolled to suspended matter would hold good in a fully charged stream flowing with the critical velocity it is not easy to say.

As silt deposits in the pond, the velocity of the water in it, along the course of the main current towards the canal, increases and eventually the water begins to carry coarse sand dangerous for the canal. In order to ascertain when this state of affairs has been reached, two methods of procedure are possible. One is to frequently test specimens of the water in the pond along the course of the main current and see when it contains more than $\frac{1}{15,000}$ of its volume of coarse sand. This plan would be troublesome and liable to error, and is rejected by Kennedy, who suggests that the depth and velocity of the water in the pond be frequently observed along the course of the main current. As soon as the velocity exceeds the critical velocity for mixed silt, it is time to close the canal and open the under-sluices and scour out the deposit from the pond. The period in which most silt is believed to have been deposited in the canal is the spring and early summer, say from 15th March to 1st July. This is the time when the snows are melting and the river water is clear. It can then carry more sand than in the rains—1st July to 15th September,—when it is muddy.

Kennedy also suggests that some under-sluices should be provided at the far side of the river, *i.e.* at the right-hand side of the weir. It would then be possible, by opening them, to let floods pass without interfering with the pond.

The two spurs or groynes, shown in the plan, were constructed in 1897 so as to cause the stream to flow along the face of the canal regulator and not allow deposits to accumulate there. The depth of silt deposited in a great part of the pond amounted at times to 8 or 10 feet.

6. **Protection of the Bed.**—It is possible to afford direct protection from scour to the bed of a stream by constructing walls across it, but unless the walls are near together the protection will not be effective. An arrangement used in some streams in Switzerland consists of tree trunks secured by short piles and resting on

brushwood. But as long as the walls are not raised above the bed they cannot entirely stop scour, unless extremely close together. If raised above the bed they form a series of weirs.

The weirs must be so designed that the depth of water in a reach between two weirs is great enough to reduce the velocity down to the critical velocity, or less. The fall in the water surface at each weir being very small, the discharge over the weir can be found by considering it as an orifice extending up to the downstream water surface, and the head being the fall in the surface at the weir.

To stop scour of the bed by direct protection without raising the water-level, the bed can be paved, a plan adopted in artificial channels with very high velocities. The paving can be of stones, bricks, or concrete blocks. The Villa system of protection, which has been used in Italy, France, and Spain, consists of a flexible covering laid on the bed. Prisms of burnt clay or cement are strung on several parallel galvanized iron wires, which are attached to cross-bars so as to form a grid a few feet square. The grids are loosely connected to one another at the corners, and the whole covering adjusts itself to the irregularities of the bed (*Min. Proc. Inst. C.E.*, vol. cxlvii.).

The special protection or paving required in connection with weirs and such-like works is considered in CHAP. X., *Arts. 2* and *3*.

CHAPTER VI

WORKS FOR THE PROTECTION OF BANKS

1. **Preliminary Remarks.**—The protection of a length of bank from scour may be effected by spurs, which are works projecting into the stream at intervals, or by a continuous lining of the bank. A spur forms an obstruction to the stream (CHAP. IV., *Art. 1*), and when constructed, or even partly constructed, the scour near its end may be very severe, even though there may be little contraction of the stream as a whole. If the bed is soft a hole is scoured out. Into this hole the spur keeps subsiding, and its construction, or even its maintenance, may be a matter of the greatest difficulty. A high flood may destroy it. If it does not do so, it may be because the stream has, for some reason, ceased to attack the bank at that place. A continuous lining of the bank is not open to any objection, and is generally the best method of protection. Spurs made of large numbers of rather small trees, weighted with nets filled with stones, have been used on the great shifting rivers of the Punjab which swallowed up enormous quantities of materials. The use of spurs on such rivers has now, in most cases, been given up. If L is the length of a spur measured at right angles to the bank, the length of bank which it protects is about 7 L—3 L upstream and

4 L downstream,—but the spur has to be strongly built, and its cost is, in many cases, not much less than the expense of protecting the whole bank with a continuous lining.

Whatever method is adopted, a plan, large enough to show all irregularities, should always be prepared, and the line to which it is intended that the bank shall be brought marked on it.

Sometimes natural spurs exist as, for instance, where a tree projects into a stream or has fallen into it, and the holes between the spurs may be deep, so that a continuous

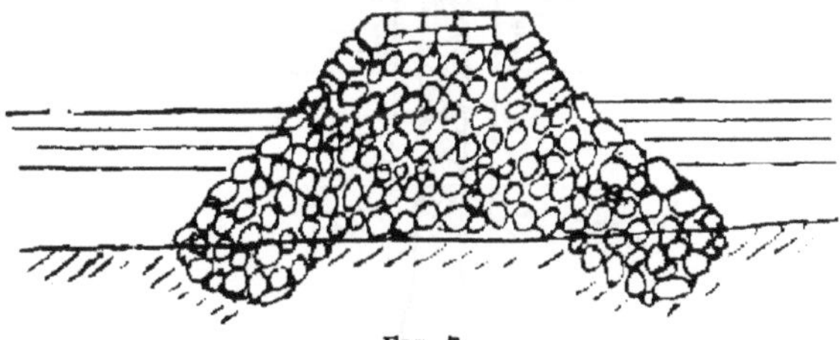

Fig. 7.

protection would be expensive. Or there may be trees standing in such positions that, if felled, they will be in good places for spurs. In cases such as the above, spurs may be suitable even in a stream with a soft channel.

Regarding the use of spurs or groynes for diversion works or for reducing the width of a stream, see Chap. VII., *Art. 1*; and Chap. VIII., *Art. 3*.

2. **Spurs.**—A spur may be made of—

(*a*) Loose stone, which may be faced with rubble above low-water level (fig. 7).

(*b*) Layers of fascines weighted with gravel or stones.

(*c*) Earth or sand closely covered with fascines.

(*d*) A double line of stakes with fascines or brushwood laid between them (fig. 8).

(*e*) A single line of stakes with planking or basket work on its upstream side, or with twigs or wattle laid horizontally and passed in and out of the stakes, as in fig. 20.

(*f*) A single tree with the thick end of the trunk on the bank and with stakes, if necessary, to prevent the current from moving it.

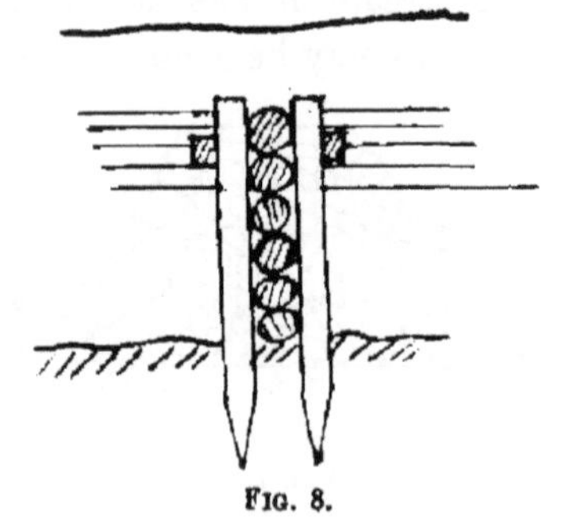

FIG. 8.

(*g*) A number of small trees heaped together and weighted with nets full of stones.

(*h*) A layer of poles and over them a layer of fascines on which are built walls of fascine work so arranged as to form cells or hollow rectangular spaces which become filled with silt.

(*i*) Large fascines running out into the stream and having their inner ends staked to the bank while the outer ends float, other fascines being added over them and projecting further into the stream, and the whole eventually sinking.

Combinations of the above are also used, for instance,

(*d*) or (*e*) may be used for the upper portion, the foundation being (*a*) or (*c*).

Instead of running out at right angles to the bank a spur may be inclined somewhat downstream. This

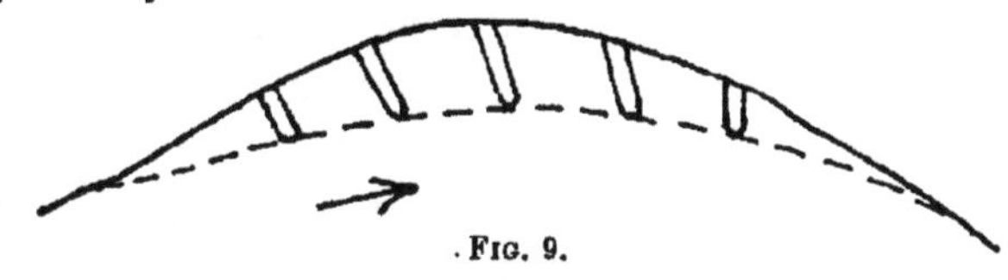

FIG. 9.

somewhat reduces the eddying and scour round the end. The ends of a system of spurs should be in the line which it is intended that the edge of the stream shall have (fig. 9). The tops of short spurs are usually above

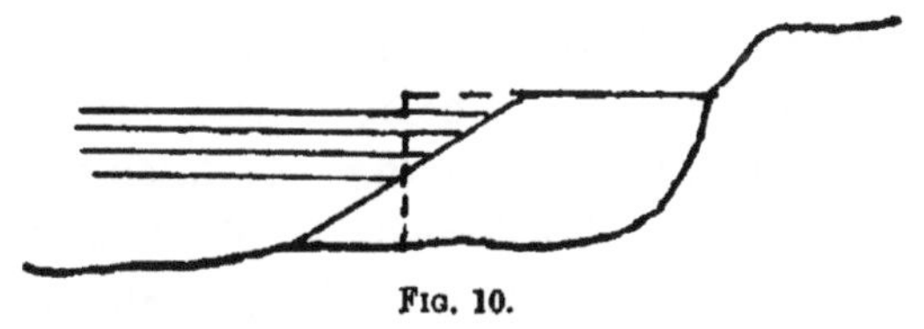

FIG. 10.

high flood level. Sometimes spurs are made to slope downwards (fig. 10), and they then cause less disturbance of the water and less scour than if built to the form shown by the dotted line. Such spurs are sometimes combined with a low wall running across the bed of the stream,

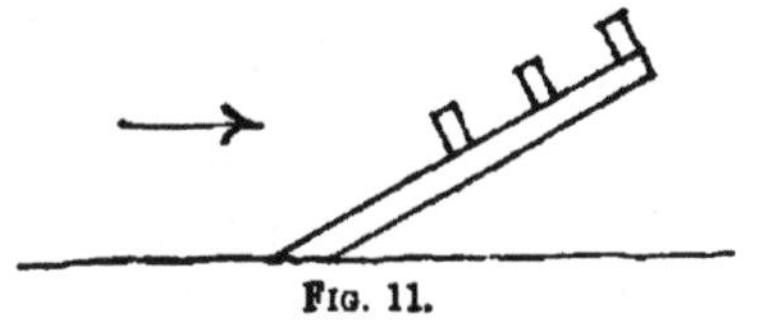

FIG. 11.

the whole forming a "profile" of the cross-section to which it is intended to bring the channel. Regarding such walls, see CHAP. V., *Art. 6*. When a spur is long it may have small subsidiary spurs (fig. 11) to reduce the rush of water along it; or its end may have to be protected in the same manner as the advancing end of a closure dam (CHAP. VII., *Art. 2*).

The following is a curious case of misconception of the action of spurs. In 1909 the river Indus was eroding its right bank and threatening to destroy the town of Dera Ghazi Khan. A clump of date palms formed a promontory and resisted erosion to some extent. A suggestion was made—by an engineer of eminence who had formerly been consulted in the case—to the effect that the date palms be removed, the reason given being that they caused disturbance and scour. On this principle spurs would have to be made not to protect a bank but to cause it to be eroded.

3. **Continuous Lining of the Bank.**—The lining or protection of a bank may be of stone or brick pitching (figs. 12 and 13), loose stone (fig. 14), fascines (fig. 15), turfing, plantations, brushwood, or of other materials laid on the slopes. Before protecting a bank it is best to remove irregularities and bring it to a regular line. This can generally be done most easily by filling in hollows, but sometimes it is done by cutting off projections. It is also necessary to make the side slope uniform. Where the slope is as shown by the dotted lines in figs. 12 to 14, filling in can be effected, but cutting away the upper part of the slope is also feasible. Such cutting away has been proposed as a remedy in itself in cases where the steep upper part of the slope was falling in, but it is not much of a remedy.

Stone pitching may rest, if boats are required to come close to the bank, on a toe wall of concrete, as in fig. 13,[1] or otherwise on a foundation of loose stone, as in fig. 12. When concrete is used the bed is dredged to such a

[1] See also Appendix B.

depth as will provide against undermining by scour. Sloping boards attached to piles are placed along the front face and the concrete is thrown in under water. The slope of stone or brick pitching is usually from

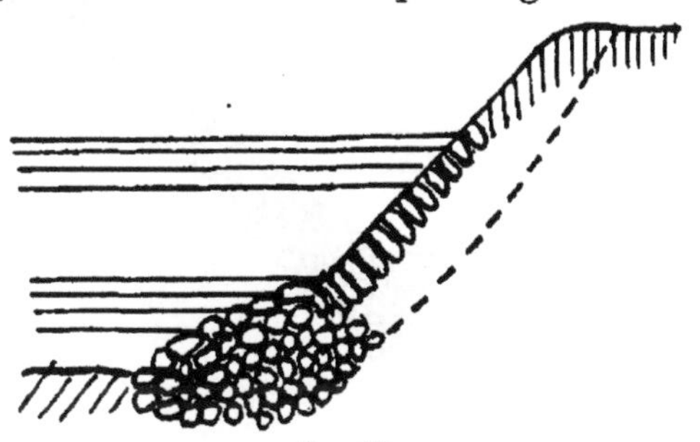

FIG. 12.

2 to 1 to 1 to 1, but it may be as steep as ½ to 1. The earth behind the pitching must be well rammed in layers. In order to prevent the earth from being eaten away by the water which penetrates through the inter-

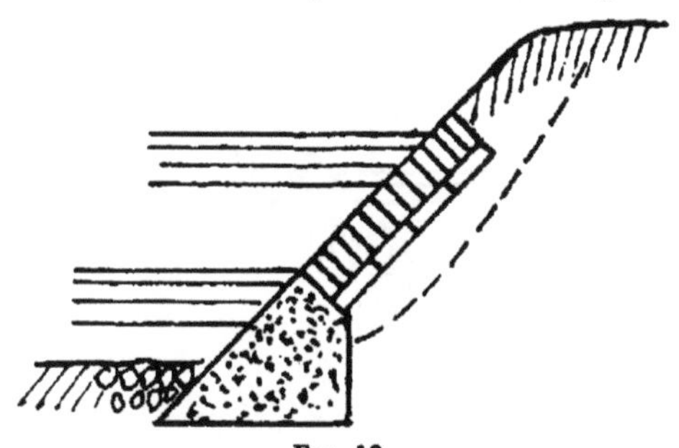

FIG. 13.

stices of the stone or brick, a layer, 3 to 6 inches thick, of gravel or ballast is placed over the earth and rammed. When loose stone is used, dredging is not necessary, but the stone is allowed to gradually sink down and more is added at the top. A certain proportion of the stones should be of large size.

When fascining is used, long twigs are made into bundles and tied up at every 2 feet so as to form fascines about 4 to 6 inches thick, and these are laid

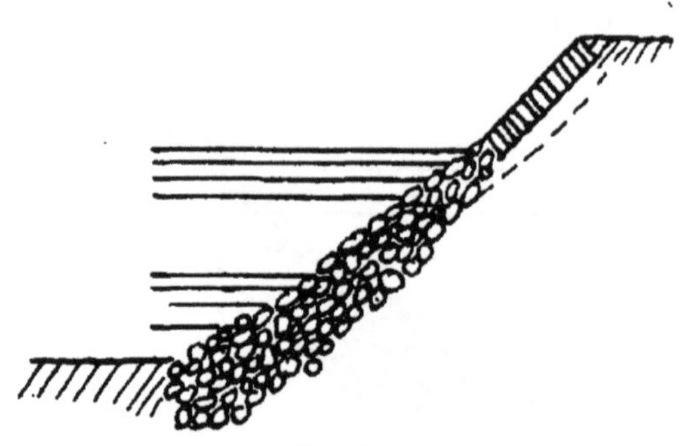

FIG. 14.

on the slopes and secured by pegs driven in at short intervals, between the fascines.

Sometimes the pitching or loose stone is not carried up to the top of the bank, or even up to high flood-level,

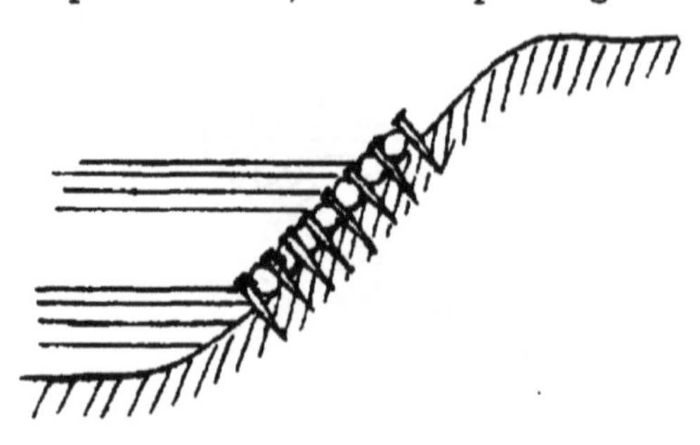

FIG. 15.

and the bank above the pitching is protected by turfing —the pieces of turf being placed on edge normally to the slope if very steep (fig. 14) or laid parallel to the slope if it is not very steep—or, above ordinary water-level, by plantations of osiers or willows which obstruct the water and tend to cause silting, and whose roots bind the banks together.

Another method of using fascines is to lay them on the slopes with their lengths normal to the direction of the stream. The upper end of a fascine is above low

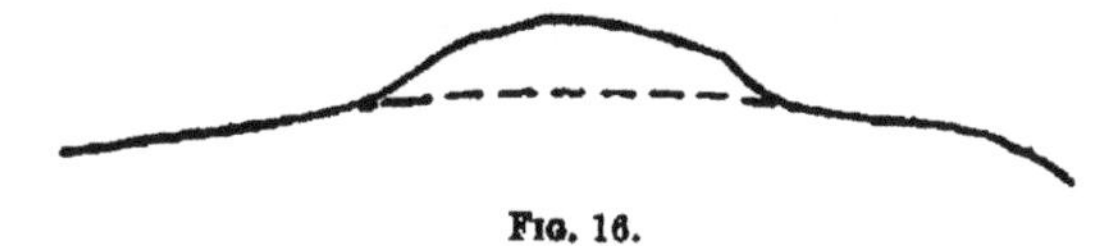

FIG. 16.

water, and the lower end extends down to the bed of the stream. Sometimes large ropes made of straw, or rough mats made of grass, are laid on the slopes and pegged down, or mattresses of fascines are laid on the slopes and weighted with stones.

A deep recess in the bank (fig. 16) can be filled in, before the protection is added, with earth well rammed.

FIG. 17.

On the Adige the filling material consisted (fig. 17) of faggots filled with stones, small cross dams being made at intervals, as shown by the dotted lines, to arrest flood water and cause it to deposit silt. At the back of the berm, poplar or willow slips were planted, and these grew up and their roots held the bank together. This system succeeded well.

A method of protection which is suitable when the

water contains much silt is what is known in India as bushing. Large leafy branches of trees are cut and hung, as shown in fig. 18, by ropes to pegs. They must be closely packed so as not to shake. At first they require looking after, but silt rapidly deposits and the branches become fixed and no longer dependent on the

FIG. 18.

ropes. If the work is carefully done, the result is a smooth, regular, and tenacious berm, as per dotted line in the figure.

Another method, used on canals, is to make up the bank with earth and to revet it with twigs or reeds, as shown in fig. 19. The foundation must be taken down

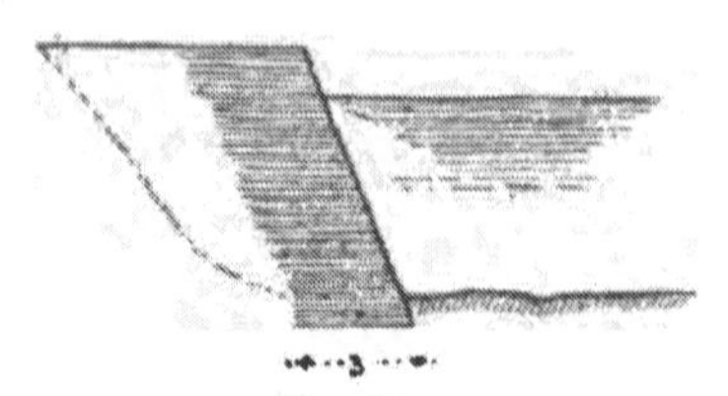

FIG. 19.

well below bed-level, otherwise the work may slip. This kind of work cannot be done except when the canal is dry.

If the bank consists of sand or of very sandy soil, it must in any case have a flat side slope such as 3 to 1. If the sand is in layers alternating with firm soil, it is a good plan to dig out some of the sand and to replace it with clods of hard earth.

Staking (fig. 20) may be used, the stakes being one or two feet apart from centre to centre, and long twigs laid horizontally being passed in and out of the stakes, or bushing filled in behind the stakes. But bushing alone is cheaper and nearly as good.

For protecting the banks of the Indus it has been proposed (*Punjab Rivers and Works*, CHAP. IV.) to use trees in exactly the same manner as bushing, the trees being grown in several rows parallel to the river so that whenever the river, by eroding its bank, comes up to the lines of trees the first row will fall in. The first row

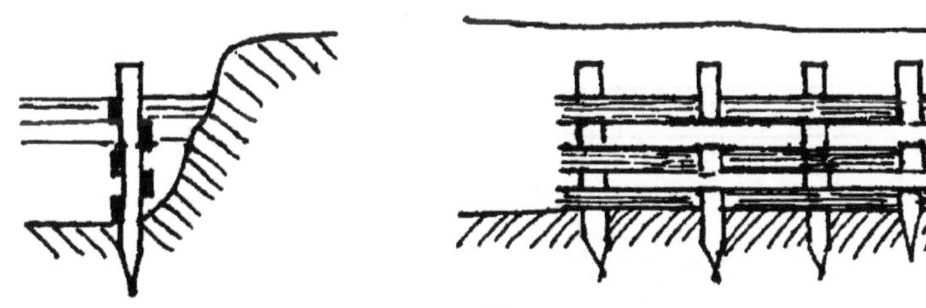

FIG. 20.

would be chained to the second, which would take the place of the pegs used in bushing. The other rows would remain as a reserve.

The Villa system of bed protection (CHAP. V., *Art. 6*) has also been successfully used for bank protection on the Scheldt, and on the Brussels-Ghent Canal, the prisms being about $10 \times 10 \times 4$ inches, and having overlapping joints. The bands of prisms are placed in position by a boat, the bands unrolling over a drum. The boat is provided with an oscillating platform carrying rollers at its end. A thin layer of gravel is laid over the bank and is pressed down by the rollers before the prisms are laid on it (*Min. Proc. Inst. C.E.*, vol. cxxxiv., and vol. clxxv.).

In the case of the river mentioned in CHAP. XI., *Art. 3*, where extremely high velocities were met with, cylindrical rolls of wire-netting were made, each 50 feet long and 5 feet in diameter, and filled with boulders. These rolls can be used for bank protection. The netting was made by wires 6 inches apart, crossing each other at right angles and tied together at the crossings by short pieces of wire.

On ship canals a berm (fig. 21) is frequently made a few feet below the water-level. It serves as a foundation for the pitching, which need not usually extend down to

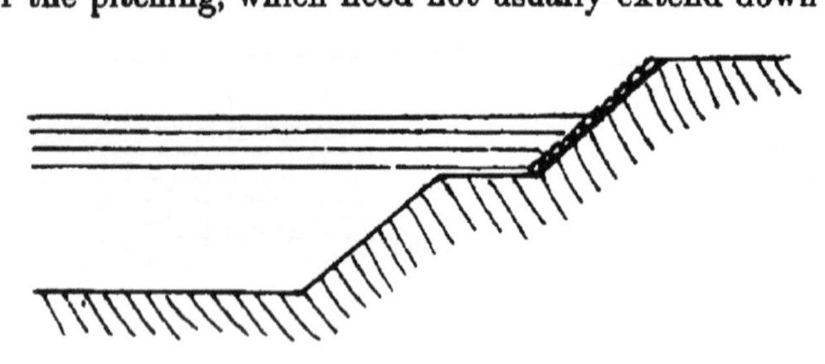

FIG. 21.

more than 5 feet below the water-level. Below that the wash has little or no effect on the banks. On ordinary navigation canals a similar berm is sometimes made—one or two feet in width and a foot or less below the water-level—and rushes are planted on it.

Sometimes a bank has been protected by a kind of artificial weed, consisting of bushes or branches of trees attached to ropes. The end of the rope is fastened to the bank and the weeds float in the stream alongside the bank.

To protect a bank from ice, which exercises an uplifting force on pitching, use has been made of a covering of a kind of reinforced concrete consisting of slabs of concrete with wires embedded in it, and fastened to the bank by wires, 20 inches long, running into the bank, these wires being embedded in mortar so as to act like stakes.

4. **Heavy Stone Pitching with Apron.**—On the great shifting rivers of India a system of bank protection is adopted, consisting of a pitched slope with an apron (fig. 22). The system is used chiefly in connection with railway bridges or weirs, but it has been used in one instance, that of Dera Ghazi Khan, for the protection of the bank near a town. When, as is usual, the flood-level is higher than the river bank, an artificial

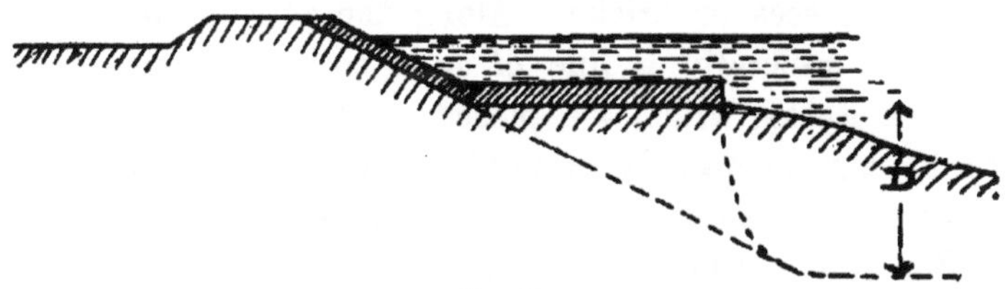

FIG. 22.

bank is made. In any case the bank is properly aligned. The pitching has a slope of 2 to 1, and consists of quarried blocks of stone loosely laid, the largest blocks weighing perhaps 120 lbs. The apron is laid at the time of low water on the sandbank or bed of the stream. If necessary, the ground is specially levelled for it. It is intended to slip when scour occurs. The following dimensions of the apron are given by Spring (*Government of India Technical Paper*, No. 153, "River Training and Control on the Guide Bank System," 1904). The probable maximum depth of scour can be calculated as explained in CHAP. XI., *Art. 3.* If this depth, measured from the toe of the slope pitching is D, and if T is the thickness considered necessary for the slope pitching, then the width of the apron should be 1·5 D, and its thickness 1·25 T next the slope and 2·8 T next the river. It will then be able to cover the scoured slope to a thickness of 1·25 T. This thickness is made greater than T because the stone is not likely to slip quite regularly. The thickness T should, according to Spring, be 16 inches to 52 inches, being least with a slow current and a channel of coarse sand, and greatest with a more rapid current and fine sand; but since the sand is generally finer as the current is slower, it would appear that a thickness of about 3 feet would generally be suitable. Under the rough stone there should be smaller pieces or bricks. Along the top of the bank there is generally a line of rails so that stone from reserve stacks, which are placed at intervals along the bank, can be quickly brought to the spot in case the river anywhere damages the pitched slope.

For the special protection to banks required near weirs and similar works, see CHAP. X., *Arts. 2* and *3*.

CHAPTER VII

DIVERSIONS AND CLOSURES OF STREAMS

1. **Diversions.** — When a stream is permanently diverted the new course is generally shorter than the old one, and the diversion is then often called a cut-off. The first result of a cut-off is a lowering of the water-level upstream and a tendency to scour there, and to silt downstream of the cut-off. Fig. 23 shows the longi-

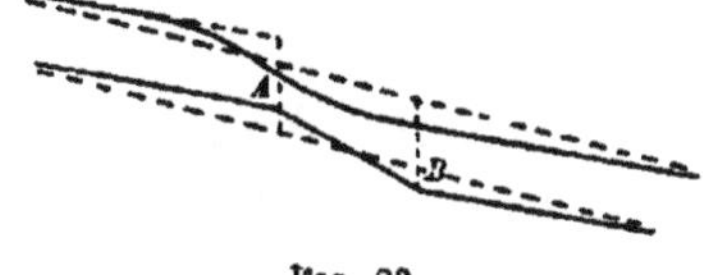

FIG. 23.

tudinal section of a stream after a cut-off *A B* has been made. The bed tends to assume the position shown by the dotted line. If both the diversion and the old channel are to remain open, the water-level at the bifurcation will be lowered still more, and the tendency to scour in the diversion will be reduced.

If the material is soft enough to be scoured by the stream, it is often practicable to excavate a diversion to a small section and to let it enlarge itself by scour. This operation is immensely facilitated if the old channel can be closed at the bifurcation. The question whether the scoured material will deposit in the channel downstream of the diversion must be taken into consideration ; also the question whether the diversion will continue to enlarge itself more than is desirable. The velocity in the diversion will be a maximum if its section is of the "best form," *i.e.* if its bed and sides are tangents to a semicircle whose diameter coincides with the water surface, but this may not (CHAP. IV., *Art. 6*) be the section which will give most scour. In order to prevent the enlargement of the diversion taking place irregularly, the excavation can be made as shown in fig. 24, water being admitted only to the central gullet. The side gullets should not be quite continuous, but unexcavated portions should be left at intervals, so that if the water

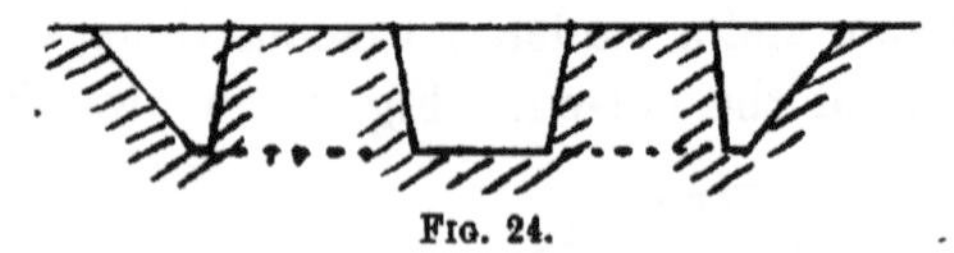

FIG. 24.

in scouring out the channel breaks into the gullet, it will not be able to flow along it until it has broken in all along.

If a diversion is made, not with the object of lowering the water-level but merely in order to shorten the channel, the increased velocity caused by the steepened slope may be inconvenient. In this case a weir or weirs can be added (CHAP. VIII., *Art. 4*).

If the water contains sufficient silt to enable the abandoned loop to be silted up within a reasonable time, it may be desirable to do this. The silting up may, for instance, increase the value of the land. The loop should be closed at its upper end. Water entering the lower end will cause a deposit there. When the lower end is well obstructed by silt, the upper end should be opened.

The set of the stream, due, for instance, to a bend at the point where a diversion takes off has very little to do with the quantity of water which goes down the diversion. The only effect of the set of the stream is a slight rise of the water-level as compared with the opposite bank. Similarly, the angle at which the diversion takes off is only of importance in giving, in some cases, a velocity of approach whose effect is generally small. The distribution of the water between the diversion and the old channel really depends on their relative discharging capacities. If the required quantity of water does not flow down a diversion it can be dredged.

Sometimes a long spur is run out to send the water towards the off-take of a diversion. The effect of this is very small—it merely causes a set of the stream,—unless its length is so great that it amounts to something like a closure dam. It is sometimes said that it is easier to lead a river than to drive it. This remark is probably based on the fact that spurs, such as those under consideration, generally produce little effect, whereas the excavation of a diversion or the deepening of a branch by dredging it, is more likely to produce some result. There is, however, no certainty about this. Sometimes too much is expected of such channels. Calculations are not always made as to the scouring power of the stream, nor is account always taken of the fact that as the cut scours its gradient flattens.

2. **Closure of a Flowing Stream.**—The closure of a flowing stream by means of a dam is usually attended with some difficulty and sometimes with enormous difficulty. There may be little trouble in running out dams from both banks for a certain distance, but as soon

as the gap between the dams becomes much less than the original width of the stream, the water on the upstream side is headed up and there is a rush of water through the gap, which tends to deeply scour the bed and to undermine the dams. The smaller the gap becomes the greater is the rush and scour.

The closure is most easily effected at or near to the place where the stream bifurcates from another. Then, as the gap decreases in width, some of the water is driven down the other stream and it does not rise so much. Eventually all the water goes down the other stream, and the total rise is only so much as will enable this other stream to carry the increased discharge. If the closure is not effected near a bifurcation, the rise of the water will go on even after the closure is completed, and it will not cease, unless the water escapes or breaks out somewhere, until it has risen to the same level as that to which it would have risen if the closure had been at the bifurcation, or perhaps not quite to the same level, since there may still be a slight slope in the water surface and a small discharge which percolates through the dam. Sometimes in such a case it is possible to arrange for temporary escapes or bifurcations, which will be shallow and therefore easily closed, after the main closure has been completed.

A closure is, of course, far more easily effected where the bed is hard than where it is soft. Very often it is best to close temporarily at such a place or near a bifurcation, even if the permanent dam has to be elsewhere, and then to construct the permanent dam in the dry channel, or in the still water, and remove the temporary one or cut a gap in it.

Generally the best method to adopt in a closure is to

cover the bed of the channel beforehand—unless it is already hard enough—with a mattress or floor, such that it cannot be scoured as the gap closes. A floor may consist of a number of stones or sandbags dropped in from boats or by any suitable means, and placed with care so that there shall not be gaps or mounds. Sandbags should be carefully sewn up. A mattress may be made of fascines laid side by side and tied together, floated into position, weighted and sunk. Even a carpet made of matting or cloth and suitably weighted has sufficed in some cases. If the scour is likely to be such that stones or sandbags will be carried away, the stones may be placed in nets, baskets, or crates. Sandbags may also be placed in nets. Probably the long rolls of wire-netting filled with stones, described in CHAP. VI., *Art. 3*, would be better than anything, and the diameter could be reduced somewhat. The floor or mattress need not usually extend right across the stream. It must cover a width much greater than—perhaps twice as great as—the width of the gap is likely to be when scour begins. Its length, measured parallel to the direction of the stream, must be such that severe eddies in the contracted stream will have ceased before the stream reaches its downstream edge. It need not extend to any considerable distance upstream of the line of the dam.

The dams when started from the banks can generally be of simple earth or gravel, or loose stones, but before they have advanced far they will probably require protection at the ends by stones, or by staking and brushwood, or by fascines. As soon as the dams have advanced well onto the mattress and their ends have been well protected, it is best to cease contracting the stream from the sides and to contract it from the bottom by laying a number of sandbags across the gap so as to form a submerged weir. In this way the rush of water is spread over a considerable width of the stream. The weir is then raised until it comes up above water. Leakage can be stopped by throwing in earth, or gravel, or bundles of grass on the upstream side. Sometimes it is best to construct the mattress over the whole width of the stream, and to effect the closure entirely by a weir, carrying each layer right across before adding another. The banks of the stream, if not hard, can be protected by sandbags, stones, staking or fascining.

The chief cause of failures of attempts to close flowing streams is neglect to provide a proper floor or mattress. The stones or other materials may be of insufficient weight or not closely laid, or the extent of the floor may be insufficient. In a soft channel and deep water loose stones in almost any quantity may fail unless a mattress of fascines is laid under them. Another cause of failure is running short of materials, such as sandbags. Allowance should be made for every contingency, including making good any failure of parts of the work. Enormous sums of money have been wasted, and vast inconvenience, loss and trouble incurred, in futile attempts to close breaches in banks, or gaps in dams.

Sometimes the gap is closed by sinking a barge loaded with stones, or by sinking a "cradle" or large mattress made of fascines, taken out to the site by four boats, one supporting each corner, and then loaded with stones and sunk. Another method is to run out a floating mattress of fascines from one side of the gap to the middle and sink it, then to proceed similarly on the other side, and so on.

An excellent plan, when it can be adopted, is to have more than one line of operations, so that the heading up of the water is divided between them.

In India closures of streams having depths of 6 or 8 feet are effected by means of rough trestles made from trunks of small trees and placed at intervals in the stream like bridge piers, one leg of the trestle inclined upstream and one downstream. Each pair of adjacent trestles is connected by a number of rough, horizontal poles. Against these are placed bundles of brushwood. Earth is at the same time collected and is rapidly added at the last. The chief danger is the undermining of the bed by scour. This is prevented by driving in stakes and placing brushwood against them. Closures of small channels or of breaches in the banks of canals are effected by means of staking and brushwood. Where dangerous breaches are liable to occur, it is a good plan to have a barge, fitted up with a small pile-driver and carrying a supply of sheet piles, ready at a convenient spot.

Hurdle dykes, first used on the Mississippi, were employed on the Indus in 1902 to close partially the main channel of the river. There were to be three dykes, each dyke consisting of three lines of very long piles—some were 60 feet long,—driven into the bed of the stream, which was to be protected with mattresses made of fascines and extending right across it, with their heads above flood-level. The idea was not to wholly stop the flow of the water, but to obstruct it so much that silt would deposit, the channel become choked up, and the water find a course down another channel. The work was begun in March 1902, and was in progress in May of the same year when an unusually early flood put a stop to it. The dykes had at this time advanced considerable distances from the right bank of the stream, but none had been completed. Two dykes out of the three were for the most part carried away. The river, however, took a new course, starting from a point far upstream, the western channel became a creek, and the remains of the dykes were soon embedded in silt.

In any case in which the provision of a proper mattress has been omitted, or when the mattress has been destroyed, or when a breach has occurred in an embankment, whenever, in short, it is evident that the gap cannot be closed until some other escape for the water is provided, it may be possible to provide such an escape by cutting partly through the dam or embankment on the downstream side at another place, and thoroughly protecting the place and extending the protection downstream and away from the dam or embankment. The water can then be let in, and the closure of the old gap attempted. If a closure is effected, the protected gap can then be closed. Sometimes it may be desirable to make such a protected gap beforehand and with deliberation.

Dams for closing streams which are dry can be made similarly to flood embankments (CHAP. XII., *Art.* 7). Sand does very well, provided it is protected by a covering of clay or by fascining.

3. **Instances of Closures of Streams.**—In 1904 the Colorado River broke into the Salton Sink—a valley covering 4000 square miles. Unsuccessful attempts

were made to close the stream by two rows of piles with willows and sandbags between them, by a gate 200 feet long, supported on 500 piles, and by twelve gates each 12 feet wide. A "rock-fill" dam was then constructed on a mattress 100 feet wide and 1·5 feet thick. The river, which was 600 feet wide, broke through, but was stopped by the construction of three

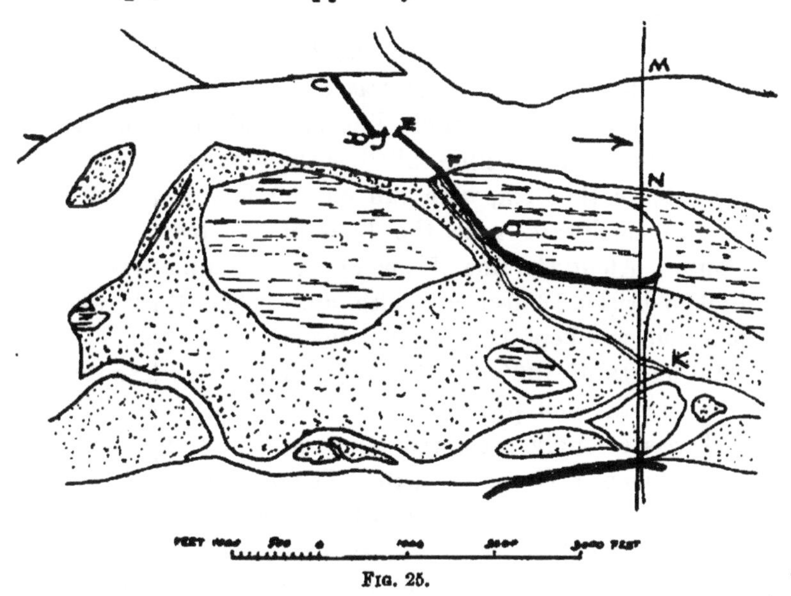

FIG. 25.

parallel rock-fill dams in the gap (*Min. Proc. Inst. C.E.*, vol. clxxi.).

At the site of the railway bridge over the river Tista in Bengal, it was necessary to close the main stream (fig. 25), which flowed at the left side of the channel, while the bridge had been built at the right. The bed was of sand, width 500 feet, depth 6 feet, and discharge 3700 cubic feet per second. The first attempt to close the stream was made at M N, a floor of stone 200 feet long, 20 feet wide, and 2 feet thick, being laid in the middle of the stream, and dams of earth, sandbags, and stones being run out from each bank. As the gap decreased in width the bed was torn up and the work failed. The heading up was 3 feet 9 inches. It was recognised later that the site should have been at the bifurcation higher up, and that the stone floor should have been laid on a mattress.

In the next working season the dams C D and E F G were made. The dam C D was of earth. Two walls, each consisting of a double line of bamboos with the spaces between the lines filled with bundles of grass weighted with earth, were run out 50 feet in advance of the earthwork near the lines of the toes of the slopes. Along the line of the upper wall a mattress of broken bricks 10 feet in width, and 1 foot thick, was laid, and was kept 50 feet in advance of the wall. A total length of 1000 feet of embankment was made in five months and pitched on its upstream side. The end was strongly protected by a mass of stone. The embankment F G was of earth. The dam E F consisted of three lines of piles driven 10 feet into the bed. A mattress weighted with stones extended for 20 feet upstream of the dam and 40 feet downstream. A gap of 150 feet was left at D E, and was not protected by a floor of any kind. A channel, parallel to F G and extending to K, had been dug to a width of 200 feet. During the floods the heading up at D E was about 2·5 feet, and the water was 30 feet deep. The line E F was greatly damaged and was repaired. The cut F G K gradually enlarged, and by the end of the floods more water was going down it than down the main stream. The gap D E was finally closed by means of a line of bamboos and grass, the bed being protected by a carpet, 100 × 50 feet, made of common cloth weighted with sand-bags. The success of the operations turned on the scouring out of the cut F G K. It is remarkable that the gap D E did not become wholly unmanageable in the floods (*Min. Proc. Inst. C.E.*, vol. cl.).

CHAPTER VIII

THE TRAINING AND CANALISATION OF RIVERS

1. **Preliminary Remarks.**—When a stream is trained or regularised it is generally made narrower, but sometimes narrow places have to be widened. Deepening has also very frequently to be effected. The object of training is generally the improvement of navigation, but it may be the prevention of silt deposit. Some natural arms of rivers which form the head reaches of canals in the Punjab are wide and tortuous, and they are sometimes trained. Training often includes straightening or the cutting-off of bends, as to which reference may be made to CHAP. VII.

2. **Dredging and Excavating.**—When a flowing stream is to be deepened, the work is usually done by dredgers. Dredgers can remove mud, sand, clay, boulders, or broken pieces of rock. The "bucket ladder" dredger is the commonest type. The "dipper" dredger is another. Both these can work in depths of water ranging up to 35 feet. The "grab bucket" dredger can work up to any depth and in a confined space. The "suction dredger" draws up mud or sand mixed with water. A dredger may be fitted with a hopper or movable bottom, by means of which it can discharge the dredged material—this, however, involves cessation of

work while the dredger makes a journey to the place where the material is to be deposited—or it can discharge into hopper barges or directly on to the shore by means of long shoots. For small works in comparatively shallow water the "bag and spoon" dredger, worked by two men, can be used.

When rock has to be removed under water it is blasted or broken up by the blows of heavy rams provided with steel-pointed cutters.

In widening a channel the excavation can be carried down in the ordinary way to below the water-level, a narrow piece of earth, like a wall, being left to keep the water out. If the channel cannot be laid dry, the work can be finished by dredging.

Regarding methods by which the stream is itself made to deepen or widen its channel, reference may be made to CHAP. V.

3. **Reduction of Width.**—If a channel which is to be narrowed is not a wide one, the reduction in width can be effected by any of the processes described under bank protection (CHAP. VI.). But in a wide channel, reduction of the width by any direct process is generally impracticable. The expense would generally be prohibitive. Earth, if filled in, is liable to be washed away unless protected all along. Reduction in the width of a large channel is nearly always effected either by groynes (fig. 26) or by training walls (fig. 27). Spurs or short groynes for bank protection have been already described (CHAP. VI., *Art. 2*). Groynes for narrowing streams are made in the same way and of the same materials, but are longer. They are at right angles to the stream or nearly so. Groynes in the river Sutlej have been mentioned in CHAP. V., *Art. 5*, and are shown

in fig. 6, p. 55. Whether groynes or training walls are used, the object is to confine the stream to a definite zone and to silt up the spaces at the sides. These spaces when partly silted can be planted with osiers or

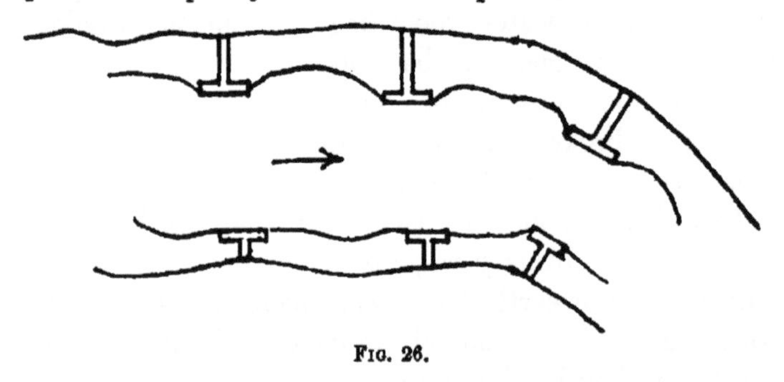

FIG. 26.

with anything which will grow when partly submerged, and this will assist in completing the silting.

A training wall can be made of any of the materials used for groynes. In order to silt up the spaces between

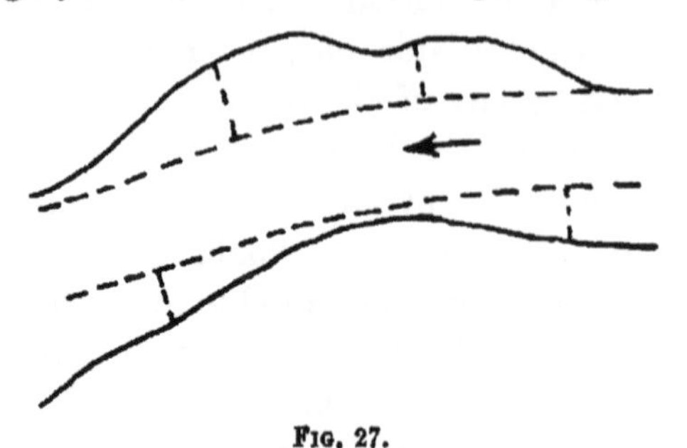

FIG. 27.

each wall and the adjacent bank of the stream, other walls are run at intervals across them. Usually the training walls and cross walls are carried up only to ordinary water-level, sometimes only to low-water level. Floods can thus spread out and submerge the walls and deposit silt. If the walls are carried up too high it may be necessary, in order to give room for floods, to space them too far apart, and this, as will be seen below, is objectionable.

The difference between training walls and groynes is one of degree rather than one of kind. The material most commonly used is, in either case, loose stone—with pitching, if desired, above low-water level,—but it may be wattled stakes. If the water of the stream contains silt at all stages of the supply, gaps can be left in training walls so that silt deposit may occur at all times and not only in floods. If the walls are of wattled stakes, water will pass through them, and it may not be necessary to leave any gaps. Groynes are frequently made with T-heads (fig. 26), and they are thus equivalent to training walls with long gaps in them. The edge of the narrowed channel usually forms somewhat as shown in the figure. If the groynes are placed so near together as to give a regular channel, the cost is not likely to be much less than that of training walls.

The alignment of training walls or groynes should be such as will give the best channel consistent with economy in cost. The best channel is generally that which is most free from sharp bends. It is assumed for the present that no cuts or diversions of such lengths as to materially alter the gradient are to be made, but that a certain amount of choice of alignment is afforded by the reduced width of the trained channel and by small diversions or easings of bends. It is sometimes said that straight reaches are objectionable because the stream will tend to wander from side to side and cause shoals, whereas in a bend there will be no such tendency. The difficulty as to shoaling will be greatest at low water, but it is likely to be serious only when the width between the training walls is too great. If the width cannot be reduced to such an extent as to do away with the trouble, it may be better to adopt a curved course. The width between the training walls should generally be the same throughout, whether the reaches are straight or curved, but in view of the preceding remarks it may be desirable, where a reach cannot be otherwise than straight and where shoaling is feared, to give the straight portion a reduced width with of course a greater depth, and similarly to reduce the width at reverse changes of curvature. In curves which are at all sharp the curvature should be rather sharper in the middle of the curve than at the ends (CHAP. IV., *Art. 8*).

4. **Alteration of Depth or Water-Level.**—When the width of a stream is altered, the depth of water—the gradient being supposed to be unchanged—must alter in the opposite manner. A narrowing of the channel by training necessitates an increase in the depth of water, and the same remark applies if an arm of the stream is closed. The increase in depth may be effected either by raising the water-level or by lowering the bed—as may be convenient—or both. If the bed is to be lowered and is of hard clay, it may be necessary to dredge it and, when this has been done, training may be unnecessary. If the bed is of soft mud, a dredged channel is likely to fill up again, and training alone will be the method to adopt. If the bed is moderately hard, say compact sand, it may be suitable to train the channel first and then to dredge if necessary. In any case, shoals of hard material may have to be dredged or rocks, whether these form shoals or lateral obstruc-

tions, to be blasted or otherwise broken up (*Art. 2*). In cases where it is desired to raise the water-level without any lowering of the bed, training is of course necessary. In any case in which the bed is likely to scour to a lower level than is desired, or if the bed is to be raised, the measures described in CHAP. V., *Art. 6*, may be adopted, but they are hardly likely to be suitable and satisfactory in all cases.

5. **Training and Canalising.**—The steps so far described, together with any of those described in CHAPS. V. and VI., exhaust the list of what can be done so long as only the cross-section of a stream is

FIG. 28.

dealt with. This is often called the "regulation" of a stream, though "training" is a more satisfactory term.[1]

A mere alteration of the cross-section of a stream will not always afford a solution of the problem to be solved. Frequently a change of gradient is required. The gradient can be steepened by means of straightenings, or flattened by introducing weirs, or perhaps by adopting a course somewhat more circuitous than was intended. This extended scope of operations is known as canalising in the case of a river, and remodelling in the case of a canal.

Suppose that it is desired to alter the cross-section of a stream, at ordinary water-level, so as to reduce the width and increase the depth (fig. 28). If the mean depth is doubled, the new width will be about equal to

[1] On Indian canals the term "regulation" is applied to the control of the discharge at the regulators or off-take works.

$\frac{1}{3 \cdot 2}$ of the old width (*Hydraulics*, Chap. VI., *Art. 2*). If this gives too narrow a channel, it may be desirable to flatten the gradient. If it gives too wide a channel, the gradient can be steepened or a greater depth adopted. While the width and depth of the stream will be fixed so as to be suitable for the navigation, the ratio of depth to velocity should be so arranged, if this is possible, as to minimise trouble connected with silting or scour (Chap. IV., *Art. 6*). A remodelled channel is, in short, designed in exactly the same way as a new channel. The depth of water exercises the greatest effect on the discharge, and the gradient the least. The weak point in a scheme which includes weirs is the difficulty of dealing with floods. A scheme perfect in all other respects may be vitiated because of the obstruction, caused by weirs, to the passage of floods. The difficulty is got over by means of movable weirs. The whole subject of weirs is dealt with in Chap. X.

Training or canalising should not be effected in any reach of a stream without regard to other reaches. A mere local lowering of the water-level by dredging may accentuate the effect of a shoal at the upper end of the reach.

When the water-level is raised by a weir or by narrowing the channel—though in the latter case the raising may not be permanent—it is generally best to commence the work from the upstream end. The raising of the water-level will then not interfere with the execution of the rest of the work. But in a case of widening, where the water-level upstream of the work is lowered, the work can conveniently be begun at the downstream end, and the remark applies also to a case of straightening, provided that the new channel is not so small that it at first causes no lowering. In any case in which there is a doubt whether the whole of the scheme will be carried out, the reach to be dealt with first can be decided on according to circumstances. There is no general reason for selecting an upstream or downstream reach, except that any raising or lowering of the water-level will extend upstream of the reach and not downstream of it (CHAP. I., *Art. 4*).

Training walls and groynes, if made with stakes or fascines or any materials except stone, require careful watching and maintenance.

CHAPTER IX

CANALS AND CONDUITS

1. **Banks.**—All banks which have to hold up water should be carefully made. The earth should be deposited in layers and all clods broken up. In high banks the layers should be moistened and rammed. The dotted lines in fig. 29 show two possible courses of percolation water. The vertical height—from the

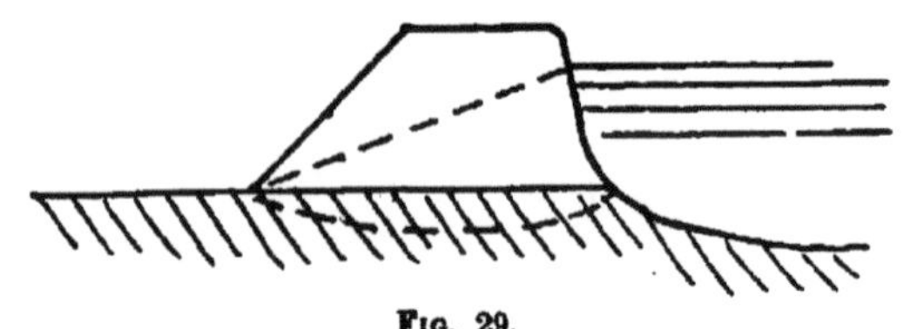

Fig. 29.

water-level to the ground outside the bank,—divided by the length of the line of percolation is the hydraulic gradient, as in the case of a pipe, and this gradient is more or less a measure of the tendency to leakage. A bank which has water constantly against it nearly always becomes almost water-tight in time. The time is less or greater according as the soil is better, and according to the amount of care with which the bank is made.

The side slopes of banks vary with the soil. Generally they are 1½ to 1, but they are sometimes 2 to 1 or even

3 to 1 if the soil is bad or sandy, or if great precautions against breaches are to be taken.

Leakage can sometimes be stopped by throwing chaff or other finely divided substances into the water at the site of the leak. In other cases it is necessary to dig up part of the bank, find the channel by which the water is escaping, and fill it up by adding earth and ramming. On some navigable canals in France it was at one time the custom to lay the reach dry, when a bad leak occurred, and to dig away the bank and lay slabs of concrete or puddle over the place. This plan was abandoned, and instead of it sheet piles are driven in. They are then withdrawn one at a time and, if any leakage occurs, the space is filled with concrete.

The dimensions of a bank should depend to some extent on the head of water against it and on the volume of the stream whose water it holds up. A breach is obviously more serious the greater the volume of the escaping water. This volume depends on the size of the stream and on its velocity. In navigation canals in England the bank on the side opposite the towing-path is usually 4 to 6 feet wide and 1½ feet above the water. In irrigation canals in India the bank of a very large canal is 2 feet above the water and 20 feet wide, while that of a small canal with 6 feet of water is 8 or 10 feet wide and 1½ feet above the water, and that of a small distributary channel with 3 feet of water is 4 feet wide and 1 foot above the water. The soil is often poor.

Further remarks, which apply to banks of special height or special importance, are given under Embankments (CHAP. XII., *Art. 6*).

2. **Navigation Canals.** — A navigation canal is sometimes all on one level, but generally different

reaches are at different levels, the change being made by means of locks. A "lateral" canal—the most common kind—runs along a river valley more or less parallel to the river. It is frequently cheaper to construct such a canal than to canalise the river. A "summit" canal crosses over a ridge and connects two valleys. A navigation canal requires a supply of water to make good the losses which occur by lockage, leakage, or absorption and evaporation. A canal may be of any size, according to the size of the boats which are to be used. There is always room, except in short reaches where the expense of construction has to be kept down, for two boats to pass one another.

A lateral canal obtains water from the river or from the small affluents which it crosses. For a summit canal it may be necessary to provide storage reservoirs. The canal crosses the ridge where it is low, and the reservoirs are made on higher ground. Reservoirs may be required also for other canals to hold water for use in dry seasons or in order to fill the canal quickly when laid dry for repairs.

In tropical countries weeds grow profusely in canals which have still or nearly still water. Traffic tends to keep them down, but they have to be cleared periodically.

In designing a barge canal the chief considerations generally are that it shall not be in such low ground or so near a river as to be liable to damage by floods, that it shall not traverse very permeable soil or gravel—this is often found near a river,— that the material excavated shall be as nearly as possible equal to that required, at the same place, for embankment, and that as far as possible high embankments, which are very expensive to construct and are more or less a source of danger, shall be avoided. The side slopes of the banks of a navigation canal depend on the nature of the soil. They are generally 1½ to 1, but the inner slope may be 2 to 1. The banks are generally 1½ or 2 feet above the water-level, the width of the bank on the towing-path side ranging from 8 to 16 feet, but being generally 12 feet and the width of the other bank 4 to 6 feet. The width of a canal is made sufficient for two boats to pass, and the depth is 1½ to 2 feet greater than the draught of the boats used. In some cases the banks are protected by pitching for short lengths, but generally they are merely turfed. The sides near the water surface wear away, so that the side slope becomes steeper in the upper part and flatter in the lower part. The resistance of a boat to traction in a canal is given by the formula

$$R = r\frac{8 \cdot 46}{2 + \frac{A}{a}},$$

where r is the resistance in a large body of water and A and a are the areas of the cross-sections of the canal and of the immersed part of the boat. When A is six times a, R is only 6 per cent. more than r. In practice A is never less than six times a.

Regarding methods of protecting banks, see Chap. VI.

A ship canal is a barge canal on a large scale. The speed of ships has to be strictly limited to avoid damage to the banks.

The Manchester Ship Canal takes in the waters of the Irwell and the Mersey, and conveys them for several miles. It is thus a canalised river for part of its course. Below that it is a tidal stream, the tide being admitted at its lower end where it joins the estuary of the Mersey, and passing out higher up where it leaves the estuary after skirting it. This circulation of water is beneficial to the estuary.

The Panama Canal might have been constructed at one level, but the cost of this, and the time occupied, would have been double that of making it a summit canal. The water of the river Chagres is to be impounded to form a lake of great extent that will not only supply water for lockage but will itself form part of the high-level reach of the canal, and ships will be able to traverse it at greater speed than in the rest of the canal.

Some Indian irrigation canals have been constructed so as to be navigable. The increase in cost has usually been enormously in excess of any resulting benefits.

3. **Locks.**—An ordinary lock is shown in fig. 29A. The space above the head gates is called the "head bay," and that below the tail gates the "tail bay." The floor of the lock is often an inverted arch. Sometimes the floor is of cast-iron. The "lift wall" is generally a horizontal arch. The gates when closed press at their lower ends against the "mitre sills"; and the vertical "mitre posts" at the edges of the gates meet and are pressed together. The gate, in opening and closing, revolves above the cylindrical "heel post"—which stands in the "hollow quoin" of the lock wall—and when fully open is contained in the "gate recess."

A lock is always strongly built, of masonry or concrete. The walls have to withstand the earth pressure when the lock is laid dry for repairs. The floor has to withstand the scouring action from the sluices. Regarding the upward pressure of the water when the lock is empty, see CHAP. X., *Art. 3.* The lift or difference in the water-levels of the two reaches of a barge canal is generally from 4 to 9 feet, but occasionally it is much more.

The gates of small locks are generally of wood and

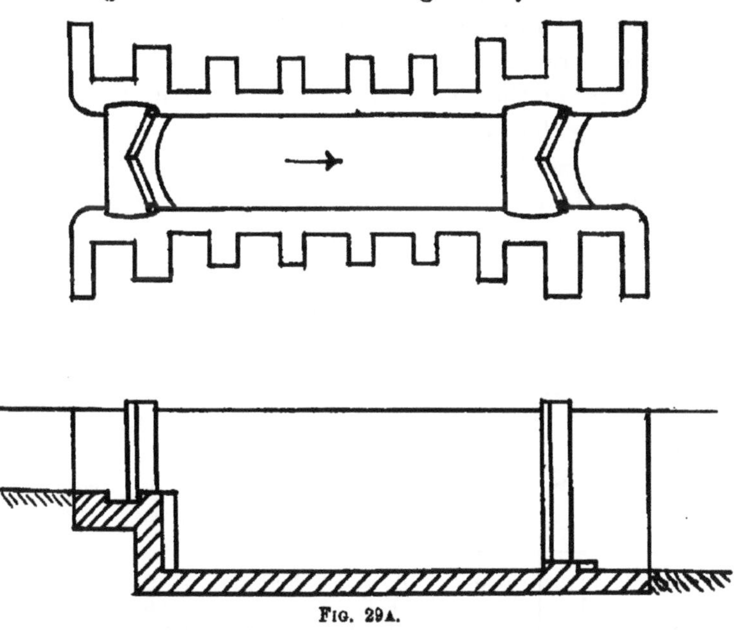

FIG. 29A.

are counterbalanced. Those of large locks are of wood or steel, and the weight is generally taken by rollers. Ordinary wood should not be used if the *Teredo navalis* exists in the waters. An iron gate, if enclosed on all sides by plating, is buoyant, and the rollers and anchor straps which hold the upper ends of the heel posts are thus relieved of much weight. The gates of the Panama Canal locks are 110 feet long and 7 feet thick, and the height ranges from 48 feet to 82 feet.

The sluices for filling and emptying a lock are placed in the gates or in the walls. The gates and sluices are generally worked by hydraulic power or by electricity.

Locks are frequently arranged in flights. There are, in a few instances, 20 to 30 locks in a flight, the total lift being 150 to 200 feet. By this means the number of gates is reduced, the tail gates of one lock being the head gates of the rest, and there is a saving in labour in working the locks.

Let L be the volume of water contained in a lock between the levels of the upper and lower reaches, and let B be the submerged volume of a boat. The "lockage" or volume of water withdrawn from the upper reach of the canal is shown in the following statement:—

Reference Number of Case.	Number of Boats.	Direction of Travel.	Lock or Locks Found	Lock or Locks Left	Lockage.	
					Single Lock.	Flight of m Locks.
1	1	Down.	Empty.	Empty.	$L-B$	$L-B$
2	1	"	Full.	"	$-B$	$-B$
3	1	Up.	Empty.	Full.	$L+B$	$mL+B$
4	1	"	Full.	"	$L+B$	$L+B$
5	$2n$	Up and down alternately.	Going down, full. Going up, empty.	Going down, empty. Going up, full.	nL	mnL
6	n	Down.	Empty.	Empty.	$nL-nB$	$nL-nB$
7	n	"	Full.	"	$(n-1)L-nB$	$(n-1)L-nB$
8	n	Up.	Empty.	Full.	$nL+nB$	$(m+n-1)L+nB$
9	n	"	Full.	"	$nL+nB$	$nL+nB$
10	n / n	Down. / Up.	"	"	$(2n-1)L$	$(m+2n-2)L$

In the case of a single lock, if two boats are to pass through, one descending and one ascending (cases 2 and 3), the descending boat would be passed through first if the lock were full, and the ascending boat first if empty; in either case, the total lockage is L, or $\frac{L}{2}$ for each boat. This also appears from case 5. Cases 6 to 10 show that if a long train of boats descends, even though the lock is full for the first boat or if a long train ascends even the lock is empty for the first boat, the total lockage is nearly L per boat. Thus in a single lock, boats should pass up and down alternately so far as this may be possible.

In the case of a flight of m locks, a single boat in descending uses no more water than if there were only one lock, the same water passing from lock to lock, but in ascending it uses more. In the case of a number $(2n)$ of boats going up and down alternately (case 5), the lockage is $m\,n\,L$, the lockage per lock per boat being $\frac{L}{2}$, but in the case of a long train of boats descending followed by an equal train ascending (cases 7 and 8), the lockage is less. If n is supposed to be equal to m, the average lockage per boat is as follows:—

m	$=1$	2	3	4	5	6	Infinity
Lockage per boat	$=\frac{L}{2}$	L	$\frac{7L}{6}$	$\frac{5L}{4}$	$\frac{13L}{10}$	$\frac{4L}{3}$	$\frac{3L}{2}$

Thus in a case where n and m are very large, the average lockage per boat, when the boats pass up and down in trains, is to the lockage per boat, when the single boats pass up and down alternately through m single locks all at different places, as 3 is to m. The reason for the difference, which may appear puzzling, is that when the locks are at different places they are worked independently of one another.

Sometimes a lock is provided with intermediate gates which provide a short lock for short vessels. In the Manchester Ship Canal, alongside each lock there is another of smaller size to be used for small vessels and thus save lockage. At the Eastham lock, where the Manchester Ship Canal descends into the estuary of the Mersey, there is, below the tail gates, an extra pair of gates opening towards the estuary, so that the lock can be worked when the water of the estuary is higher than that in the canal. Water can be economised by means of a "side-pond," into which the upper portion of the water from a lock can be discharged and utilised again when the lock has to be filled. If two locks are built side by side, each acts as a side-pond to the other. Two flights of locks can be built side by side.

Sometimes instead of a lock there is an inclined plane, up or down which are drawn on rails caissons containing water in which the boats float. The rails extend below the water-levels of the two reaches, and the caissons can thus be run under the boats. "Lifts" have also been constructed by which the boats can be lifted bodily and swung over from one reach to the other.

4. **Other Artificial Channels.**—The method of calculating the discharges of channels in which water is to flow is a question of hydraulics. The principles and rules to be followed, in the design of earthen channels, have been stated in CHAP. IV., *Art. 6*, and in CHAP. VIII., *Art. 5*. The design of banks has been dealt with in *Art. 1* of this Chapter. For conveying water for the supply of towns, or for other purposes, masonry con-

duits are often used. A usual form is shown in fig. 30. The curving of the profile of the cross-section gives an

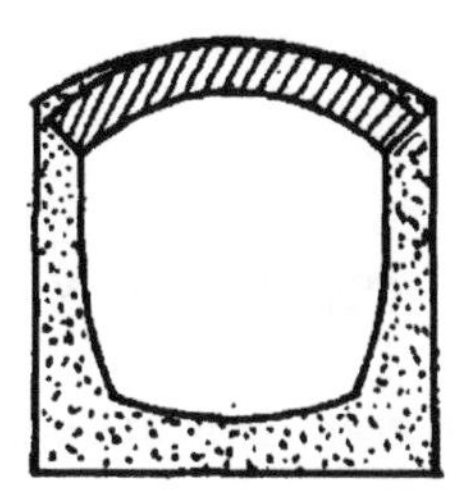

Fig. 30.

increased sectional area and hydraulic radius, and hence an increased discharge.

CHAPTER X

WEIRS AND SLUICES

1. **Preliminary Remarks.** — Every structure which interferes at all with a stream causes an abrupt change in the stream (Chap. IV., *Art. 1*). At an abrupt change there are always eddies, and these have a peculiar scouring effect. This effect is greatest where the velocity of the stream is abruptly reduced as where, for instance, after being contracted by an obstruction, it expands again or where it falls over a weir or issues from a sluice opening. In all cases of this kind the protection of the structure from scour is of primary importance.

The site of a weir or other permanent structure should, if the stream is unstable, be in a fairly straight reach, or at least not be immediately downstream of a bend. This is because of the tendency of bends to shift downstream (Chap. IV., *Art.* 8). There is no particular advantage in selecting a narrow place. A narrow place is likely to be deep or it may be liable to widen. In a hard and stable stream there is no restriction as to site.

Weirs are frequently constructed for purposes of navigation, as mentioned in Chap. VIII. They are also used in streams which are not navigable in order that the gradient may not be too steep, and in irrigation canals for the same reason. They are used both in rivers and canals in order that the water-level may be raised and water drawn off by branch channels for purposes of manufactures, water-power or irrigation.

Upstream of a weir there is more or less tendency for silt to deposit, but it by no means follows that there will be deposit (CHAP. IV., *Art. 2*, last par., and *Art. 3*, last par.). When deposit of sand or mud is feared, small horizontal passages, known as "weep holes," may be left in the weir at the level of the upstream bed. In the old Nile barrages iron gratings were provided, but they were needlessly large.

An inherent defect of an ordinary weir is that it obstructs the passage of floods. The obstruction may or may not be of consequence. Sometimes it is of great consequence. Attempts have been made to partially remedy the evil by placing the weir obliquely to the stream, thus giving it a greater length. At ordinary water-levels the flow over the crest of the weir is normal to its length, or nearly so. Supposing that the water has to be held up to a given level, the crest of the weir must be higher, because of its greater length, than if it were normal to the stream. In a flood the water has a high velocity and flows over the weir in a direction nearly parallel to the axis of the stream, so that the effective length of the weir is not much greater than if it were normal to the stream, and, its crest being higher, it obstructs the flood as much. Oblique weirs are usually made as in fig. 31. If made in one straight

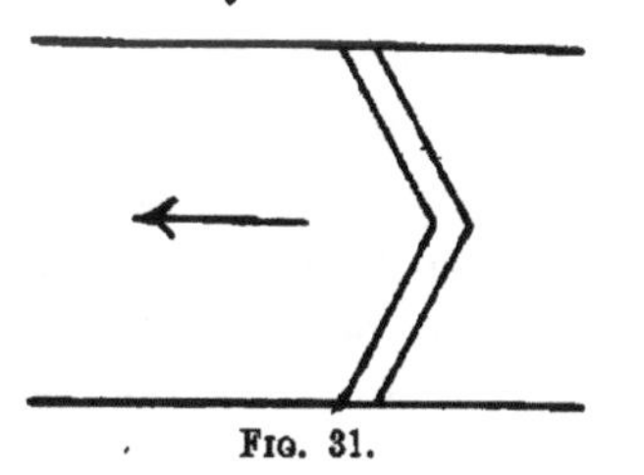

FIG. 31.

line, there might be excessive action on the bank at the lower end.

If the weir is lengthened, not by being built obliquely but by a widening of the stream at the site, the crest has to be raised and nothing is gained.

The only arrangement by which a weir can be made to hold up water when a stream is low and to let floods pass freely, consists in having part of the weir movable, *i.e.* consisting of gates, shutters or horizontal or vertical timbers, which can be withdrawn to let floods pass, and can be manipulated to any extent so as to regulate the amount of water passing. A familiar instance of a movable weir is the one which is usually placed across a mill stream, the wooden gates working in grooves in the masonry.

Above a weir in Java, 162 feet long, there was a great accumulation of shingle in the bed of the river, and the head of a canal taking off above the weir became choked. The crest of the weir on the side away from the canal was raised 5¼ feet and the crest sloped gradually down, a length of 43 feet on the side next the canal remaining as it was. This was quite successful. It was practically a contraction of the river near the canal off-take, and this must have caused scour, so that the bed became lower than the floor of the canal head and the shingle was not carried in. The shingle, however, is said to have been carried over the weir (*Min. Proc. Inst. C.E.*, vol. clxv.).

A lock is an adjunct to a weir, used when navigation has to be provided for. The lock may be placed close to the weir or it may be in a side channel, the upstream end of the lock being about in a line with the weir. Locks have already been discussed in Chap. IX., *Art. 3.*

Frequently a "salmon ladder" has to be provided. It consists of a series of steps or a zigzag arrangement so that the velocity of the water is not too great for the fish to ascend.

2. **General Design of a Weir.**—Unless the bed and sides of the channel are of rock, a weir has side walls and rests on a strong floor or "apron." These need not extend far upstream, but must extend some way downstream because of the scouring action of the water.[1] A common type of weir is shown in fig. 32. The downstream face is made sloping, so that the water

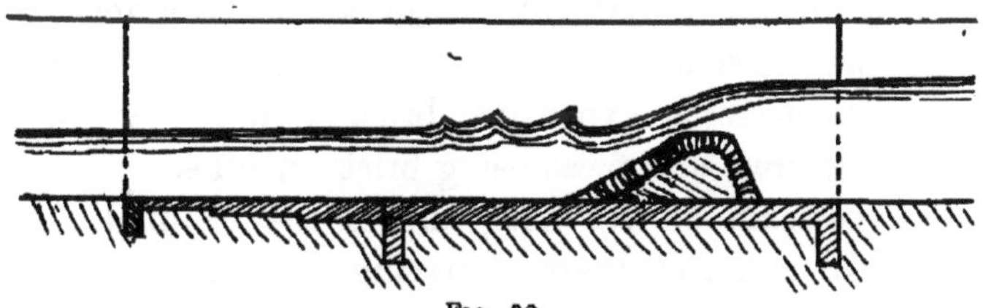

Fig. 32.

may not fall vertically and strike the floor below the weir. The thickness and length of the floor depend on the volume of water to be passed and on the height which it will fall and on the nature of the soil, and are generally matters of judgment, though rules regarding them, applicable to certain special cases, are given in the next article.

The upper corners of the weir should be rounded. This prevents their being worn away; but the rounding of the upstream corner has another advantage. If the corner is sharp, the stream springs clear from it and the weir holds up the water higher, especially in floods. With small depths of water the difference is less, and it vanishes when there is only a trickle of water. Thus a

[1] See also Appendix B.

crest rounded on the upstream side holds up low-water nearly as well as a sharp-edged crest, but lets floods pass more freely. Any batter given to the upstream face has a similar advantage. The rounding is of more importance as the batter is less. For similar reasons, the upstream wing walls should be splayed or even curved so as to be tangential to the side wall, and not built normally to the stream. These advantages are sometimes lost sight of. The downstream walls are splayed to reduce the swirl.

The body of the weir may be of rubble and the facework of dressed stone. In large weirs the stones are sometimes dowelled together. Where, as in many parts of India, stone is expensive, brick is used for small weirs, the crest and faces being brick on edge.

Downstream of the floor, unless the channel is of very hard material, there is paving or pitching of the bed and pitching of the sides, and these may terminate in a curtain wall. The bank pitching may be of any of the kinds described in CHAP. VI., *Art. 3*, and the bed paving as described in CHAP. V., *Art. 6*, but downstream of a weir the eddying is continuous and the lap of the water on the bank is ceaseless, and good methods are necessary. Sometimes planking, laid over a wooden framing or attached to piles, is used instead of paving and pitching.

In case the height of a weir is great relatively to its thickness, the danger of its being overturned must be considered. To be safe against overturning, the resultant of the pressure on the weir must pass through the middle third of its base (see fig. 62, CHAP. XIII.).

3. **Weirs on Sandy or Porous Soil.**—If the channel is very soft or sandy the weir may be built on one or

more lines of wells. The wells are not so much to support the weir as to form a curtain and prevent streams, due to the hydraulic gradient A E (fig. 33), from forming under the structure and gradually removing the soil. It is assumed in the case represented by the figure that the maximum head occurs when the downstream channel is dry. Any removal of soil from under the weir may cause its destruction. The wells should be as close together as possible, and the spaces between them carefully filled up with brickwork or concrete to as great a depth as possible, and

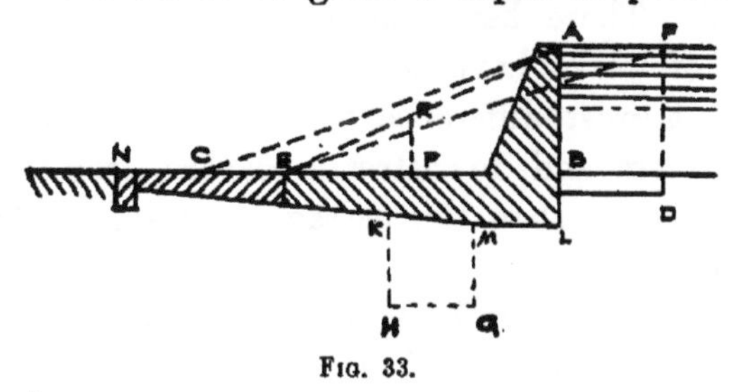

FIG. 33.

below that by piles. Instead of wells, lines of sheet piling—cast-iron or wood—can be used. A good fit should be made, but it is not necessary that the joints should be absolutely water-tight. The object is to flatten the hydraulic gradient by increasing the length travelled by the water from B E to B L G H E. Of course, no flattening occurs at a point where the curtain is not water-tight, but if only small interstices exist, none but small trickles of water can pass, and the interstices will probably soon be choked up, just as the sand in a filter bed becomes clogged and has to be washed. In any case, no important stream could develop otherwise than round the toe of the curtain. It has been stated that when a curtain is water-tight the water follows the line B L M G H K E, but this requires proof. Another plan is to cover the bed and sides of the channel with a continuous sheet of concrete extending upstream of the weir from B to D—thus flattening the hydraulic gradient from A E to F E. Instead of concrete, clay puddle can be used with pitching over it. The choice between the different methods depends largely on questions of cost and facility of construction. It has been said that a certain amount of leakage occurs under structures such as the Okla weir (*Art. 4*), which nevertheless remains undamaged. There have, however, been cases in which failures of works have occurred, especially when there has been a great difference between the water-levels of the upstream and downstream reaches, from no other apparent cause than the passage of water underneath the works.

Weirs in porous soils have been discussed by Bligh (*Engineering News*, 29th December 1910), who gives the following as safe hydraulic gradients (*s*) or ratio of the greatest head A B to the length B E:—

Fine silt and sand as in the Nile	1 in 18
Fine micaceous sand as in Colorado and Himalayan rivers	1 in 15
Ordinary coarse sand	1 in 12
Gravel and sand	1 in 9
Boulders, gravel and sand	1 in 4 to 1 in 6

These figures are probably quite safe enough even for the most important works and for those where the heading up is constant. For small works or for regulators (*Art. 5*) where the heading up is not constant, steeper gradients are permissible. Much also depends on the condition of the water. If it contains much silt, all interstices will probably become choked up. The hydraulic gradient in the case of the Narora weir across the Ganges was 1 in 11. The weir failed after working for twenty years. It was rebuilt with a gradient of 1 in 16. In the Zifta and Assiut regulators on the Nile the gradients are 1 in 16·4 and 1 in 21.

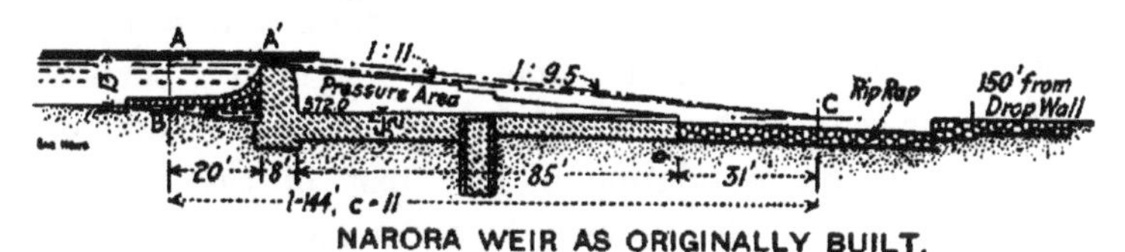

NARORA WEIR AS ORIGINALLY BUILT.

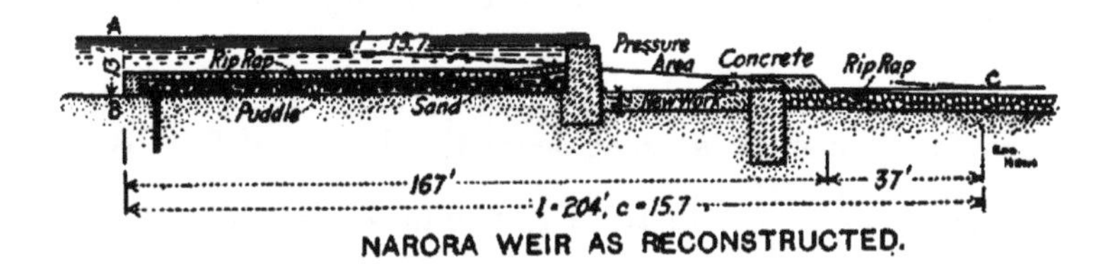

NARORA WEIR AS RECONSTRUCTED.

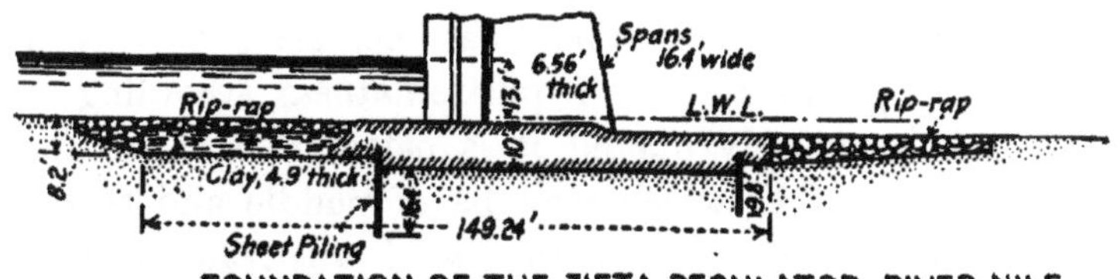

FOUNDATION OF THE ZIFTA REGULATOR, RIVER NILE.

Regarding the upward pressure on the floor due to the hydrostatic pressure from the head A B, there is a theory that the weight of a portion of the floor at any point P should be able to balance the pressure due to a head of water P R. This, supposing the masonry to be twice as heavy as water, would give a thickness of floor equal to half P R. According to Bligh, the theoretical thickness ought, for safety, to be increased by one-third. Practically the thickness need not, in most cases, be made even so great as is given by the theoretical rule. On canals in the Punjab it is certainly less. Water passing through soil or fine sand does not exert anything like the pressure which it exerts when passing through a pipe. It acts in the same manner as in a capillary tube. It is only in coarse sand or gravel or boulders that water flows as in a pipe.[1] If the tail water covers the floor, the weight of a portion of floor is reduced by the weight of an equal volume of water. If the foundation of any part of the floor is higher than B E, the upward pressure on it is reduced because the water has to force its way upwards through the soil.

Bligh also states as an empirical rule that in order to provide efficiently against scour the length of floor B E should be $\frac{4}{s}\sqrt{\frac{H}{13}}$, where H is the maximum head A B; and he points out that in a case where this length is less—as it usually is—than that necessary to give a hydraulic gradient of the requisite flatness, according to the rule previously quoted, it is better to add an upstream floor B D, which may be of puddle and therefore cheap, than to add to the downstream floor a length E C which must be of masonry or concrete, and that this arrangement, by shifting the line of hydraulic gradient from A E to F E, gives a reduced upward pressure on the downstream floor.

The length E N to which pitching, if of "rip-rap" type, should extend is given by Bligh as $\frac{10}{s}\sqrt{\frac{H}{10}}\sqrt{\frac{q}{75}}$, where q is the maximum discharge in cubic feet per

[1] *Irrigation Works*, Chap. I., *Art. 4.*

second passing over a 1-foot length of the weir, and H is the head A B.

4. **Various Types of Weirs.**—The type of weir shown in fig. 32 may be varied by steepening or flattening the slopes of one or both faces. Flattening increases the cost but gives a greater spread for the foundations. It may, however, be combined with a decrease in the width of the crest. Flattening of the downstream slope reduces the shock of the water on the floor, but the slope itself, especially the lower portion, has to stand a good deal of wear, and the

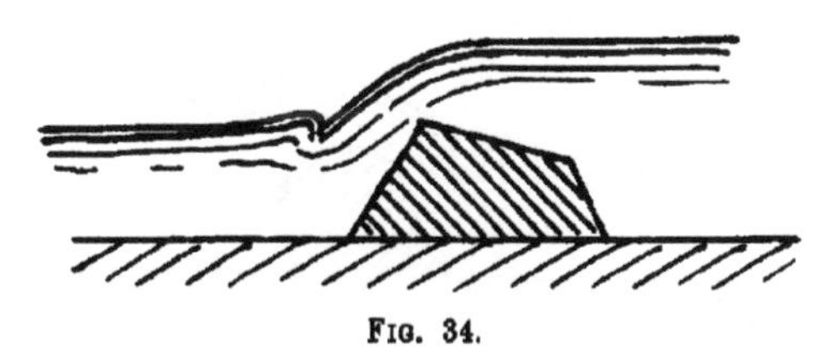

FIG. 34.

length exposed to this is increased. Flattening the upstream slope facilitates the passage of floods. The same result is obtained by making the crest slope upwards (fig. 34). In a small stream or in an irrigation distributing channel, a weir may be a simple brick wall with both faces vertical and corners rounded.

Weirs in America are often built of crib-work filled with stones. Weirs are also made of sheet piling filled in with rubble, and the top may be protected by sheet iron. A weir made on the Mersey in connection with the Manchester Ship Canal works was so made. There were three rows of piles and the filling in the back part was of clay.

Sometimes the downstream faces of weirs used to be

made curved (figs. 35 and 36), the object being to reduce the shock of the falling water, but the advantage

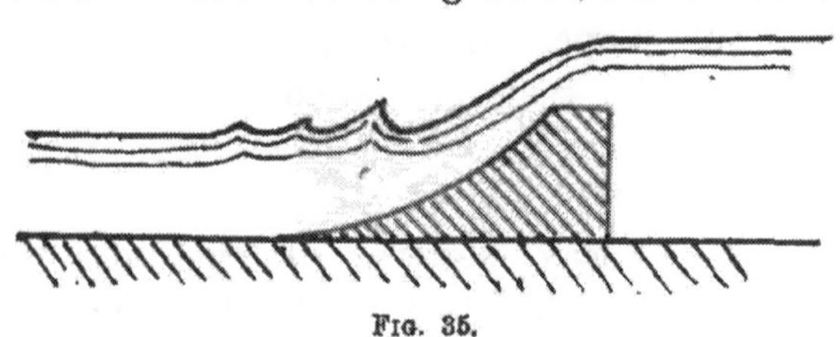

FIG. 35.

gained is not very appreciable, and this type of weir is not very common.

The Okla weir (fig. 37) across the river Jumna near

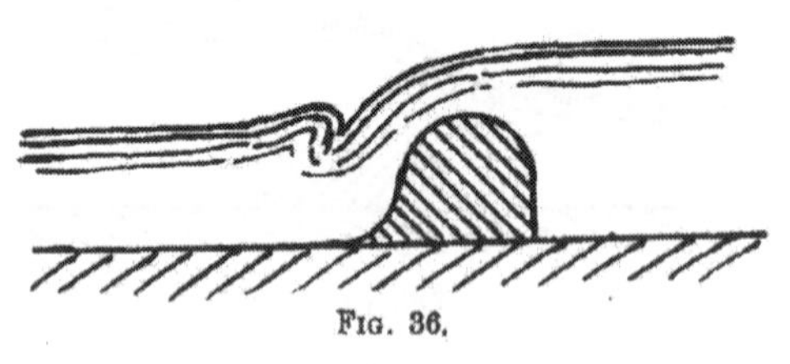

FIG. 36.

Delhi was built about thirty-eight years ago on the river bed, which consisted of fine sand. The depth of water over the crest in floods is 6 to 10 feet. The

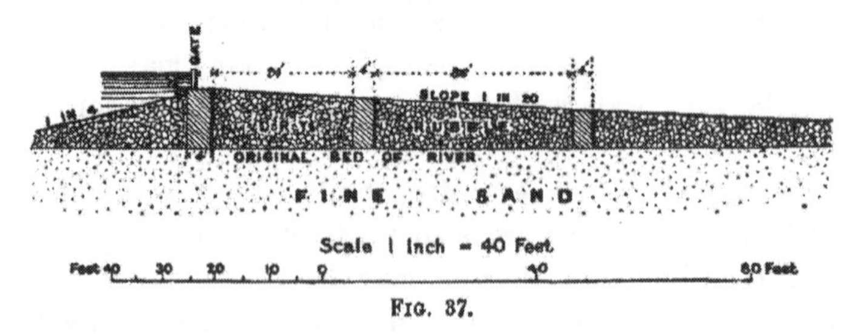

FIG. 37.

material, except the face-work and the three walls, is dry rubble.

When the reach of channel downstream of a weir has a bed-level much lower than that of the upstream reach—this is often the case in irrigation canals,—the work is known as a "fall" or "rapid." At a fall the water generally drops vertically, and a cistern (fig. 38) is provided. The falling water strikes that in the cistern and the shock on the floor is greatly reduced. An empirical rule for the depth of the cistern, measured from the bed of the downstream reach, is

$$K = H + \sqrt[3]{H}\sqrt{D},$$

where H is the depth of the crest of the fall below the upstream water-level, and D is the difference between the upstream and downstream water-levels. At some

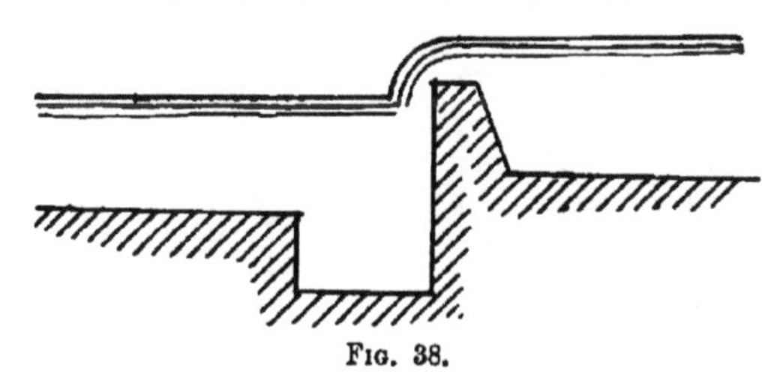

Fig. 38.

old falls on Indian canals the water, as it begins to fall into the cistern, is made to pass through a grating which projects with an upward inclination from the crest of the weir at the downstream angle. This splits up the water and reduces the shock, but rubbish is liable to collect.

In the usual modern type of canal fall in India the weir has no raised crest, and the water is held up by lateral contraction of the waterway just above the fall. The opening through which the water passes is trapezoidal (fig. 39), being wide at the water-level and narrow at the bed-level. In a small channel there is only one opening, but in a large canal there are several side by side, so that the water falls in several distinct streams.

The curved lip shown in the plan is added to make the water spread out and cause less shock to the floor. The dimensions of the openings are calculated so that however the supply in the canal may vary, there is never any heading up or drawing down. The detailed method of calculation for finding C F and the ratio of A B to B C is given in *Hydraulics*, CHAP. IV. In cases where it is only necessary for the notch to be accurate when the depth of water ranges from B C to three-fourths B C, it will suffice to calculate as follows :— Let b be the bed width of the canal, and let Q be the discharge and B the mean width of the stream when

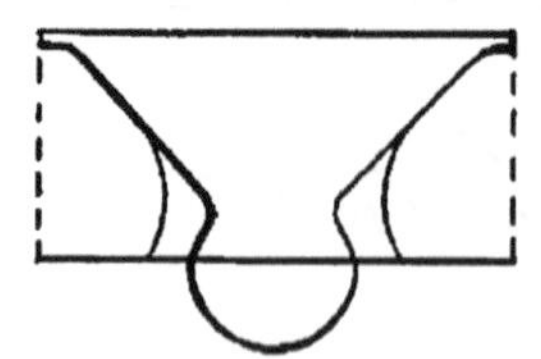

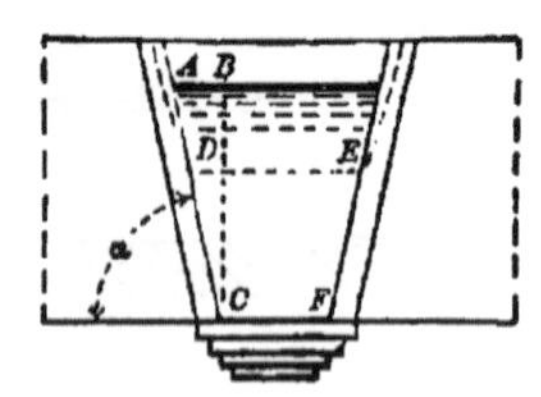

FIG. 39.

the depth of water is B C. Decide on the number of notches, and let W be the width of a notch calculated as if it were to be rectangular, *i.e.* by the ordinary weir formula. Increase the width to $W' = 1{\cdot}05\ W$. Then make the notch trapezoidal, keeping the mean width W′, and making the bottom width w (or C F), such that $\frac{w}{W'} = \frac{b}{B}$. The top width of the notch is of course increased as much as the bottom width is reduced.

A rapid has a long downstream slope, which is expensive to construct and difficult to keep in repair, especially as the canals can only be closed for short periods. Rapids exist in large numbers on the Bari Doab Canal in India, the face-work consisting in many cases of rounded undressed boulders—with the interstices filled up by spawls and concrete—which stand the wear well. Rapids have again been used on the more modern canals in places where boulders are obtainable, and where deep foundations would have given trouble in unwatering. The upstream face of a rapid is vertical, or has a steep slope.

5. **Weirs with Sluices.**—The long weirs built across Indian rivers below the heads of irrigation canals generally extend across the greater part of the river bed. In the remaining part—generally the part nearest the canal head—there is, instead of the weir, a set of

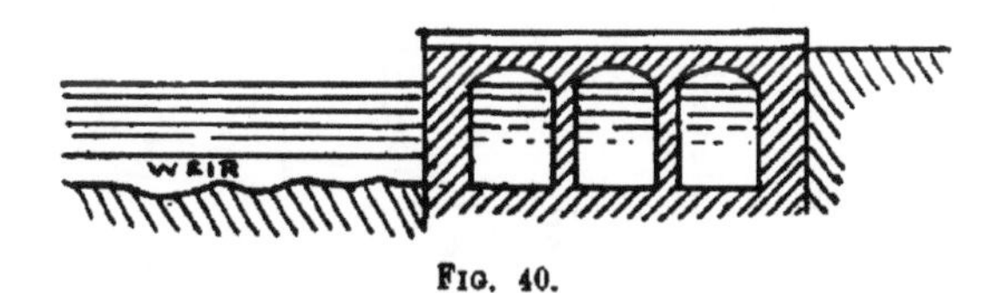

Fig. 40.

openings or "under-sluices" (fig. 40) with piers having iron grooves in which gates can slide vertically. The piers may be twenty feet apart and five feet thick. The gates are worked by one or more "travellers," which run on rails on the arched roadway. The traveller is provided with screw gearing to start a gate which sticks. When once started it is easily lifted by the ordinary gears. The gates descend by their own weight. The gate in each opening is usually in two halves, upper and lower, each in its own grooves, and both can be lifted clear of the floods. In intermediate stages of the river these gates have to be worked a good deal. (See also Chap. V., *Art. 5.*) Usually the weir has, all along its crest, a set of hinged shutters, which lie flat at all seasons, except that of low water in the river.

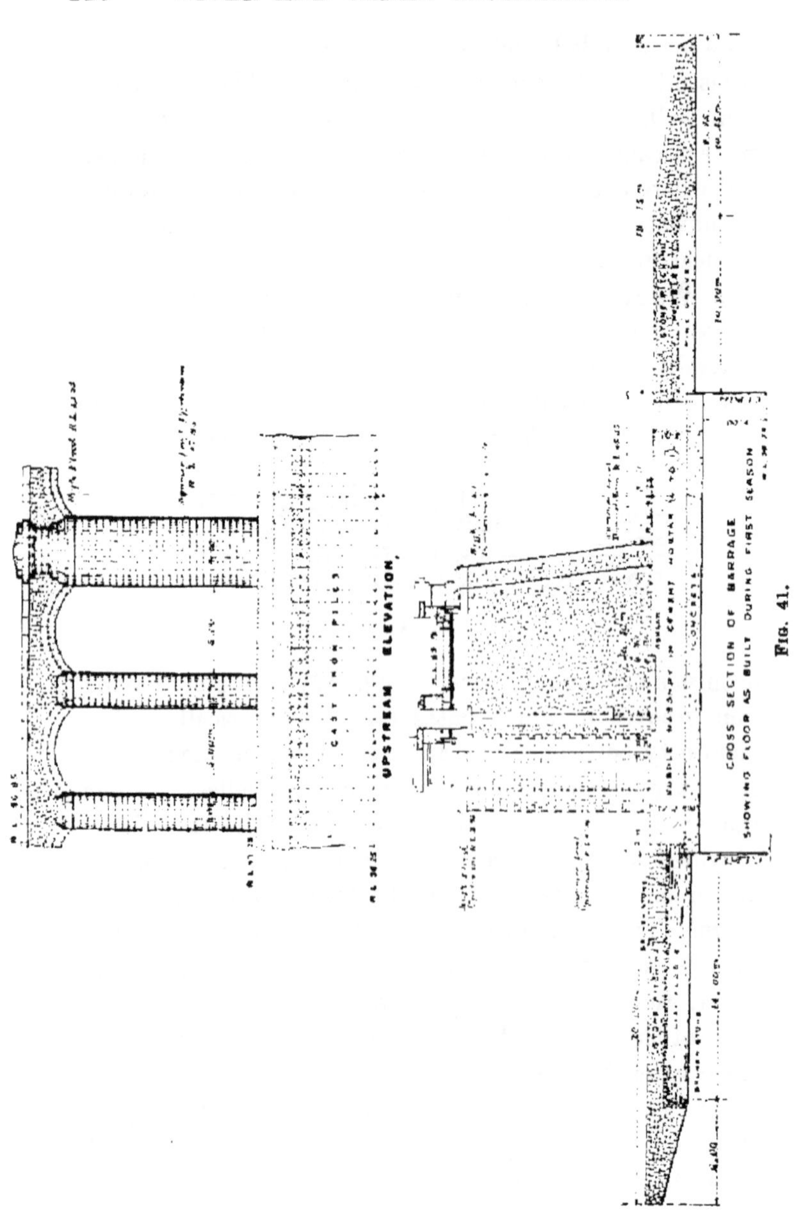
UPSTREAM ELEVATION.
CAST IRON PILES
CROSS SECTION OF BARRAGE
SHOWING FLOOR AS BUILT DURING FIRST SEASON
CONCRETE

Fig. 41.

The canal head consists of smaller arched openings, provided with gates working in vertical grooves and lifted by a light traveller. If the floor of the canal head is higher than the beds of the river and the canal, it may be said to be a weir, but otherwise the canal head is merely a set of sluices without a weir.

The barrage of the Nile at Assiut (fig. 41), and the old barrages of the Rosetta and Damietta branches, consist of sets of sluices without weirs. At Assiut there are piers five metres apart and gates working in grooves like those, above described, at Indian headworks.

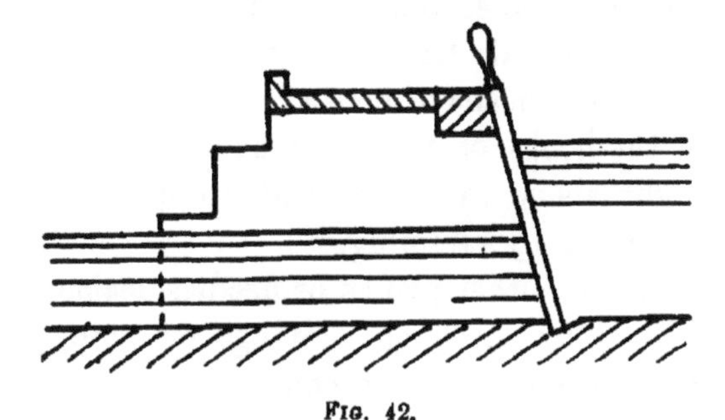

Fig. 42.

The "dam" across the Ravi, at the head of the Sidhnai Canal in the Punjab, also consists of sluice openings without a weir. The piers are connected by horizontal beams (fig. 42), against which, and against a sill at their lower ends, rest a number of nearly vertical timber "needles," fitting close together, which can be removed when necessary by men standing on a foot-bridge. In floods the needles are all removed and laid on the high-level bridge (not shown in the drawing), the foot-bridge being then submerged. With needles the span between two piers can be greater than would be possible with a gate. Needles can be used up to a length of 12 or 14 feet, excluding the handle which projects above the horizontal beam. They can be of pine, about 5 inches deep in the direction of the stream, and 4 inches thick.

Where a branch takes off from a canal in India there are usually no fixed weirs but two sets of piers—one in the canal and one in the branch,—with openings and gates like those at the canal heads, or else with wider openings and needles. These works are called regulators. The gates are worked by travellers or by fixed windlasses or racks and pinions. Very small gates for distributaries are often worked entirely by screw gearing. For the smaller branches the gates are replaced by sets of planks or timbers lying one above another and removed by means of hooks. They are replaced by means of the hooks or by being held in position some little height above the water, and dropped. They are finally closed up by ramming.

In the case of either planks or needles, leakage can be much reduced by throwing shavings or chopped straw into the water upstream of them.

Needles can be provided on their downstream sides with eye-bolts just above the level of the beam against which their upper ends rest. They can then be attached by chains or cords to the beam or to the next pier, and cannot be lost when released. They can be released by a lever which can be inserted under the eye-bolt. By pushing the head of a needle forward and inserting a piece of wood under it, a little water can be let through. In this way, or by removing needles here and there, the discharge can be adjusted with exactness.

At a needle weir in an Indian canal all the needles in one opening are reported to have broken simultaneously. A possible explanation is that one needle broke and that the velocity thus set up in the approaching stream caused the others to break. On another occasion when a canal was dry all the needles were blown down.

Sometimes the beam or bar against which the upper ends of the needles rest is itself movable. At Ravenna, in Italy, the bar between any two piers has a vertical pivot at one pier and can swing horizontally. Its other end is held by a prolongation of the next bar, near to its pivot. If the end bar of the weir is released, each bar in turn is released automatically.

At Teddington on the Thames the oblique weir, 480 feet long, has thirty-five gates, which extend over half the length of the weir. They are worked by travellers which run on a foot-bridge. The openings do not extend down to the river bed, but are placed on the top of a low weir. The other half of the weir is fixed. The gates are raised to let floods pass.

At Richmond on the Thames the arrangements are similar, the gates being counterbalanced to admit of easy and rapid raising. When raised they are tilted into a horizontal position so as not to obstruct the view.

In Stoney's sluice gates a set of rollers is interposed between the gate and the groove. The rollers are suspended from a chain, one end of which is attached to the top of the gate and the other end to the groove. The rollers thus move up or down at half the rate of the gate, and some of them are always in the proper position for taking the pressure. Escape of water between the gate and the groove is prevented by a rod which is suspended on the upstream side of the gate close to its end, and is pressed by the water against the pier. Stoney's sluice gates, with spans ranging up to 30 feet, have been used on the Manchester Ship Canal for the sluices by which the water of the river Weaver is passed across the canal, and at locks for passing the flood waters of the Irwell and Mersey down the canal. The gates are balanced by counterweights.

Frame weirs,[1] used chiefly on rivers in France but also in Belgium and Germany, are a modification of the needle and plank arrangements above described. For the masonry piers there are substituted iron frames or trestles, which are hinged at the floor-level so that, when the timbers have been removed, the frame can be turned over sideways and lie flat on the floor, thus leaving the waterway absolutely clear from side to side of the stream. The foot-bridge which rests on the frames is removed piece by piece. The frames are raised again by means of chains attached to them. In order that the frames may not be too heavy they are spaced 3 to 4 feet apart, or very much nearer than when masonry piers are used. Horizontal planks can thus be used of shorter lengths than the needles, and they can be made up into greater widths so that the leakage is less.

A further modification consists in placing the bridge platform above flood-level, and in hinging the frames to it instead of to the floor. The frame turns about a horizontal axis parallel to the length of the weir. A weir of this kind can be used for greater depths of water than the ordinary frame weir.

In some cases the horizontal planks are connected together by hinges so that they form a "curtain." The curtain is raised by rolling it up by means of a traveller. It admits of rapid and accurate adjustment of the water-level, but there is considerable scouring action below a curtain when it is somewhat raised.

[1] *Min. Proc. Inst. C.E.*, vols. lx. and lxxxv.

6. **Falling Shutters.**—In Thénard's system, first used in France, a shutter (fig. 43) is hinged at its lower edge and is held up by a strut. When the lower end of the strut is pushed aside it slides downstream and the shutter falls flat. To enable the shutter to be raised again an upstream shutter, which ordinarily lies flat and is held down by a bolt, is released, and it is then raised by the current to the extent permitted by a chain attached to it. The downstream shutter is then raised. Thénard's system was not much used in France because the river had to fall to a level somewhat too low for navigation before the shutters could be raised.

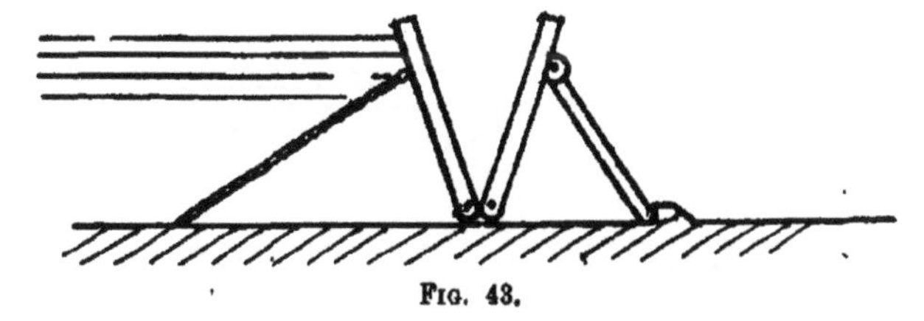

FIG. 43.

The sudden jerk on the chain of the upstream shutter is also liable to do damage. The system has been adopted on some of the long weirs which cross Indian rivers downstream of the heads of irrigation canals. To prevent damage by shock, a hydraulic brake was designed by Fouracres. It consists of a piston which travels along a cylinder and drives water out through small holes. The shutters are placed on the top of the fixed weir, where they usually lie flat, except in the low water season, any adjustments of the river discharge being effected by means of the under-sluices.

In the Chanoine system of falling shutters (fig. 44), used first in France, the shutter is hinged at a point rather higher than the centre of pressure. The hinge is supported by a vertical trestle, which is hinged at its lower end and is supported by a strut which slides in a groove and rests against a stop. When the water rises to a certain height above the top of the shutter, it is turned by the force of the water into a horizontal position. The struts can then be pushed sideways out of the stops by means of a "tripping bar," which lies along the floor parallel to the line of shutters and is worked from the bank. The struts, trestles, and shutters then fall flat. To close the weir the shutters are first raised into the horizontal position which they

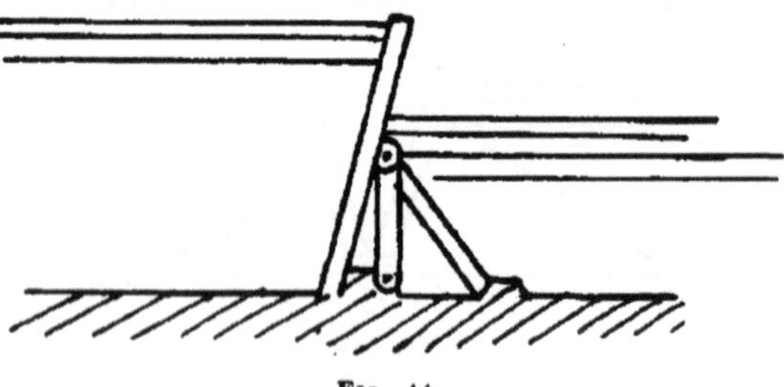

FIG. 44.

occupied before falling, by means of a hook worked from a boat or by chains attached to a foot-bridge running across the river upstream of the weir. They can then be easily closed by a boat-hook. The water closes them of itself if it falls low enough.

When the shutters fall a great rush of water occurs. To obviate this a valve is made in the upper half of the shutter. It consists of a miniature shutter on the same principle as the main shutter. The pivot of the main shutter is made at such a height that the shutter will not turn over when only a small depth of water flows over it. Instead of this the valve comes into operation. The valve also facilitates the raising of the shutter. Again, instead of the tripping bar, which would sometimes have to be of great length or be liable to damage owing to stones jamming in its teeth, the shutter can be released by pulling the strut upstream so that it falls into a second groove, down which it slides. When a tripping bar is used, its teeth can be so arranged that the shutters are released a few at a time, first singly, then in twos and threes. Sometimes there are gaps of a few inches between one shutter and the next, and the gaps can be closed by needles if necessary.

Chanoine shutters can be very rapidly lowered, and

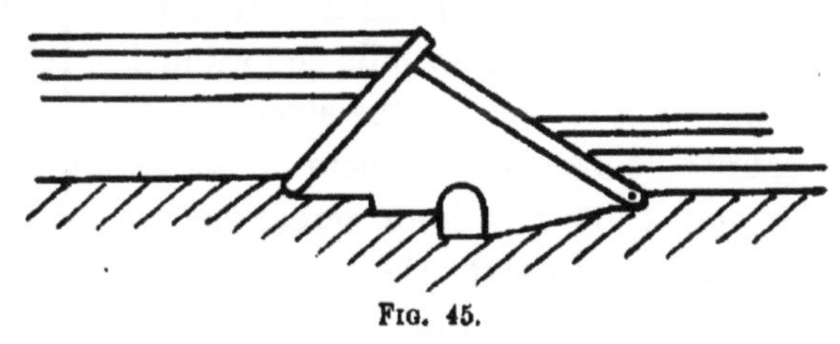

FIG. 45.

they are used in France and in the U.S.A. in places where sudden floods occur. They are also used for navigation "passes" where most of the heavy traffic is downstream and where it is too heavy to be dealt with in a lock. A foot-bridge across the stream or across the navigation pass is always an assistance, but sometimes it cannot be used when there is much floating rubbish or ice. With a foot-bridge the cost is greater than that of a needle weir.[1]

In the Bear Trap weir (fig. 45) the upstream shutter rests against the downstream one. Both are raised by admitting water from the upper reach, by means of a culvert, through an opening in the side wall, and they are made to fall by placing this opening in communica-

[1] *Rivers and Canals*, Harcourt.

tion with the downstream instead of the upstream reach. This kind of shutter is only suitable for passes of moderate width, and it is rather expensive on account of the culverts.[1]

Shutters with fixed supports are used on the Irwell and Mersey. A fixed frame is built across the stream (fig. 46) and the shutters are hinged to it. When the water rises to a certain height above its top, the shutter turns into a horizontal position, but as this causes a severe rush of water the shutter is usually raised by a chain attached to its lower end and worked from the bank. When in a horizontal position, it is held there by a ratchet. When the stream falls the ratchet is released and the shutter is closed by the stream. This kind of shutter cannot be used where there is navigation.

On the weir 4000 feet long across the river Chenab at Khanki in the Punjab, the falling shutters, 6 feet high and 3 feet wide, are hinged at the base and held up by a tie-rod on the upstream side. The trigger which releases the rod is actuated by means of a wire rope carrying a steel ball, and worked by a winch from the abutment of the weir or from one of the piers, which are 500 feet apart. A winch is fixed on the top of each pier, and communication with the piers is effected by means of a cradle slung from a steel wire rope, which rests on standards and runs across the weir. The wire rope which carries the steel ball passes over a series of forks, one on each shutter. When one trigger has been released, that shutter falls and the ball hangs loose. A further haul on the rope causes it to actuate the trigger of the next shutter, and so on. If it is desired to drop only some of the shutters, the rope is

[1] *Rivers and Canals*, Harcourt.

passed over the forks of those shutters only. The

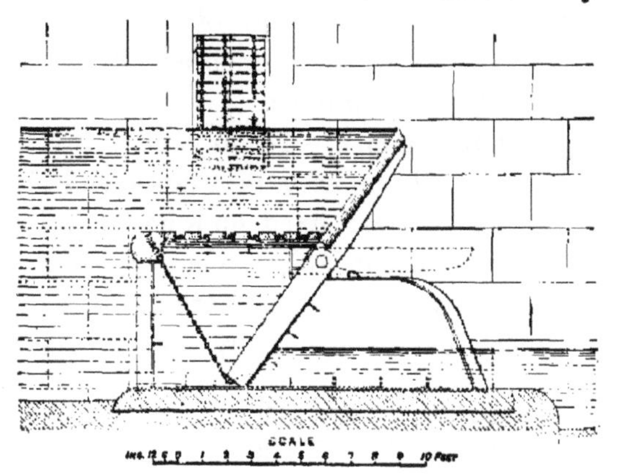

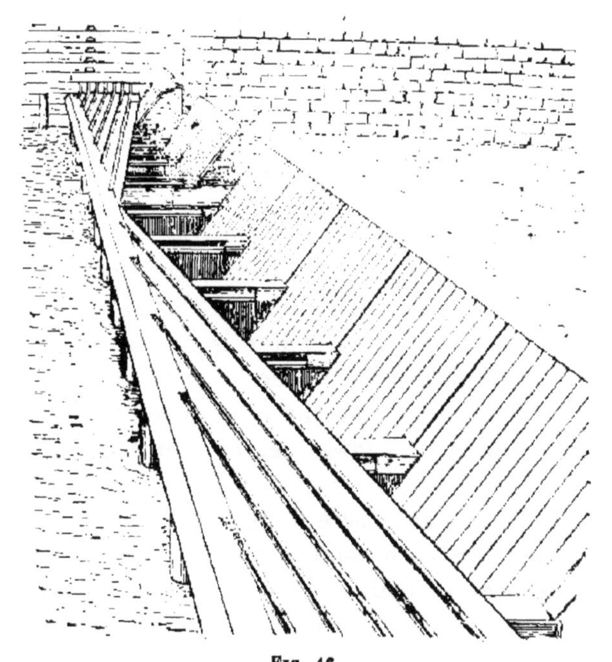

FIG. 46.

shutters can be raised by means of a crane which runs

along the weir on rails downstream of the shutters or, if the water is too high to allow of this, by a crane in the stern of a boat which is moored upstream of the weir and allowed to drop down.

7. **Adjustable Weirs.**—Drum weirs, invented by Desfontaines, have been used in France and Germany. Two paddles (fig. 47) are fixed on a horizontal axis and can turn through about 90°, the lower paddle, which should be slightly the larger, working in a "drum," which is roofed over and can, by means of sluices, be

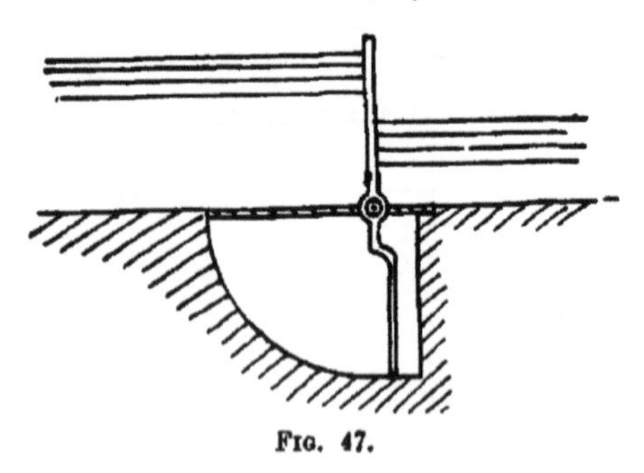

FIG. 47.

placed in communication with either the upper or lower reach of the stream. According as the upper paddle is to be raised or lowered, water is admitted from the upper reach above or below the lower paddle, the water on its other side being at the same time placed in communication with the lower reach. On the weirs first made on the Marne, the height of the upper paddle was 3 feet 7½ inches, and there were, in a weir, a number of pairs of paddles, each being 4 feet 11 inches wide. By having sluices at both abutments communicating with both reaches, and by opening or closing each of them more or less, the various paddles can be made to take up different positions, and thus perfect control over the discharge is obtained by simply turning a handle to control a sluice gate. A weir has since been made with a single pair of paddles extending right across the opening (33 feet), and the height of the upper paddle is over 9 feet.[1]

The chief objection to drum weirs is the necessity for the hollow or drum, which renders the work very expensive, except when only a small depth of water is held up.

The old sluice gates of the Nile barrages were made

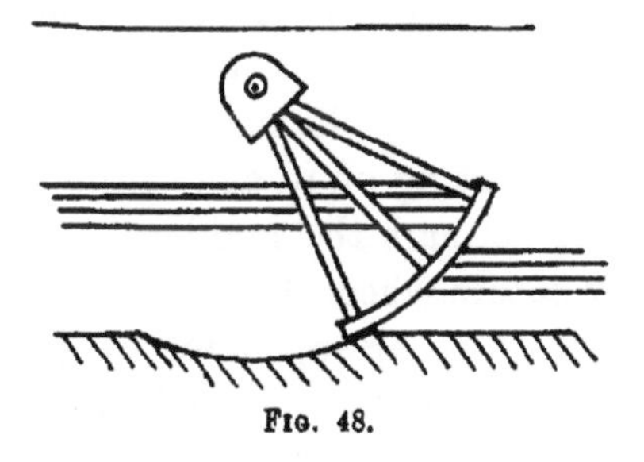

FIG. 48.

segmental (fig. 48), turned on pivots in the piers, and were raised by chains.

In some factories in Bavaria and Switzerland there are self-acting shutters which revolve on a horizontal axis at the lower edge, and are counterbalanced by cylindrical weights which roll on ways in the side wall. This arrangement is suitable when there is only one span, which can, however, be as great as 30 feet. An adjustable weir used at Schweinfurt on the Maine, consists of a hollow iron cylinder, 59 feet long and 10 feet in diameter, running across the stream. The cylinder is pear-shaped in cross-sections, and can be made, by means of mechanism, to revolve, the water passing over it. Another kind used at Mulhausen on the Rhine

[1] *Rivers and Canals*, Harcourt.

consists of a hollow iron cylinder 85 feet long and 9·8 feet in diameter. The whole cylinder can be raised by winches (*Min. Proc. Inst. C.E.*, vols. cliii. and clvi.).

8. **Remarks on Sluices.**—In all kinds of sluice openings or regulators, the principles of design as regards protection of the bed and sides, splaying and curving of walls and piers, thickness of floor, and prevention of the formation of streams under the structure are the same as laid down for weirs.

In order that a pier may be safe from being overturned by the pressure of the water when the gates or timbers are down, the resultant of its weight, including that of anything resting on it, and of the water pressure on it, must pass through the middle third of its length. This generally occurs when there is an arched roadway. Otherwise it must be arranged for by prolonging the base of the piers downstream, and giving the downstream side a batter or steps.

The floor should usually be placed at a level somewhat lower than the mean bed-level of the stream. The bed may possibly be lowered in course of time. Lowering the floor also gives a greater thickness of water cushion to take the shock of water falling over the gates or planks. It is convenient to build, on the floor, a low wall or sill, reaching up to the level of the bed or thereabouts, and running across from pier to pier under the line of gates or needles. The height of the gates or needles can thus be reduced, and there is little chance of silt or stones collecting and interfering with them. In the case of needles the wall must be strong enough to resist their horizontal pressure. If ever the bed is lowered, the wall can easily be cut down or removed.

Sluices with gates are, of course, used in connection with works other than weirs or regulators, as, for instance, in reservoirs or locks, or generally for communication between any two bodies of water. The gate may or may not be wholly submerged. If it is not wholly submerged, planks can be used. Needles can be used if the flow is always in one direction and never in the reverse direction. In all cases protection downstream of the opening is required.

In designing a set of sluice openings or regulators, it is sometimes the custom to make the total area of waterway the same as that of the stream in its unobstructed condition. There is no particular reason why it should be the same. In a description of the Assiut Barrage (*Min. Proc. Inst. C.E.*, vol. clviii., p. 30), it is mentioned that one of the reasons for placing the floor lower than the river bed was that the width of the waterway of the barrage was less than that of the river. The bed has to be heavily protected in any case, and the proper principle is to fix a velocity which is considered to be safe and, the maximum discharge being known, to determine the area of the waterway accordingly. In the case of a very wide river like the Nile, with a well-defined channel, it is inconvenient to make the distance between the abutments of a work much less than the width of the channel, but so far as velocity is concerned, the floor need not usually be lower than the bed. The protection given to the channel on the upstream side of the barrage (fig. 41) seems to be rather greater than necessary. The thickness of the floor (9 feet 10 inches) seems excessive. The thickness originally proposed was much less.

Of the many kinds of apparatus described in this

chapter each possesses some advantages and disadvantages. Gates require a bridge with powerful lifting apparatus, and are suitable for large bodies of water and great depths. Comparing needles with planks, the former can be worked by one man and admit of rapid removal, and require far fewer piers. Planks require two men, and are sometimes liable to jam, but obstruct floating rubbish less than needles, and in shallow water give rise to less leakage. Whether needles or planks are used, masonry piers are most suitable where sand or gravel are liable to accumulate on the floor, or where there is much floating rubbish. The hinged frames are suitable in other cases. Falling shutters of the Chanoine type admit of very rapid lowering, and can be used without a foot-bridge. The drum weir is perfect in action, but its cost is high.

At any system of sluices the regulation should be so arranged as to minimise the chances of damage to the bed and banks where this is at all likely to occur. If the gates are opened only near one side of the structure, there will be a rush of water on that side, and serious damage may occur. The opening should be done symmetrically and, as far as possible, distributed along the whole length.

Until experience has shown it to be unnecessary, soundings should be taken at regular periods of time downstream of every important work where scour can occur. When scour is found to have occurred at any particular part of the work, the rush of water at such places should, as far as possible, be prevented, and a chance given for silting to occur.

Unless experience shows that damage is not likely to occur, a stock of concrete blocks, sandbags, or other suitable materials should be kept on the spot ready for use. Life-buoys should be provided on any work where large volumes of water are dealt with, especially if it is unfenced in any part, or if any of the men employed are casual workers.

Regarding works for preventing a river from shifting its course so as to damage or destroy a weir or similar work, see Chap. XI., *Art. 3*.

CHAPTER XI

BRIDGES AND SYPHONS

1. **Bridges.**—Bridges are of many kinds. In this book only those parts of them are considered which are exposed to the stream. If a bridge has piers, there must be some disturbance of the water. The disturbance will be least when the area of the waterway of the bridge is at least as great as that of the stream, and when its shape is as nearly as possible the same. For small streams, a single span clearing the whole stream may be adopted, especially when the channel is of soft material, but for a large stream the cost of intermediate piers, even with a certain amount of protection for them or with deep foundations, will be more than counterbalanced by the smaller thickness of arch or depth of girder.

Generally a bridge has vertical abutments which limit the waterway, but it may have land-spans, and in this case the stream as it rises can spread out. Piers and abutments should be so designed that abrupt changes in the section of the stream are, as far as possible, avoided, the piers being rounded or boat-shaped at both ends and the abutments suitably curved (fig. 49). Boat-shaped piers, besides presenting the neatest appearance, cause the least amount of disturbance.

A bridge can be made safe against scour either by

giving deep foundations to the piers and abutments or by adding a floor and, if necessary, pitching. The former course is usually adopted and is the best. But in a case in which the discharge of a stream is to be increased or has been underestimated, it is often far easier to add a floor to an existing bridge than to increase the span of the bridge. In order to increase the waterway the floor can be "dished," *i.e.* made at a level lower than the bed of the stream[1] and gradually sloped

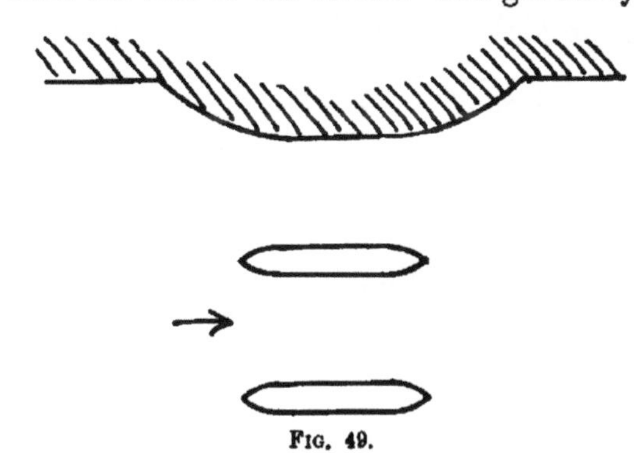

FIG. 49.

up — the slopes being pitched — both upstream and downstream of the bridge, to meet the bed.

In any case in which the water rises above the crown of the arch, the bridge becomes a syphon, and a floor is probably necessary unless the foundations are very deep, or unless the rise of water above the crown is temporary.

In the case of Indian rivers which have soft channels, and are ordinarily of moderate width but are subject to occasional floods when the width of the stream is multiplied several times and becomes very great, it is

[1] The foundations of piers and abutments should be deep enough to allow of this.

the rule to make the span of a railway bridge far less than this greater width. The stream during floods scours out a deep channel through the bridge with great rapidity, and no heading up worth mentioning occurs. The foundations of the piers are very deep, being frequently 50 feet below the lowest point of the river bed which can be found anywhere within several miles of the bridge. The span of the bridge can be arrived at by considering a general cross-section of the river as it is when in high flood, and assuming that scour to the

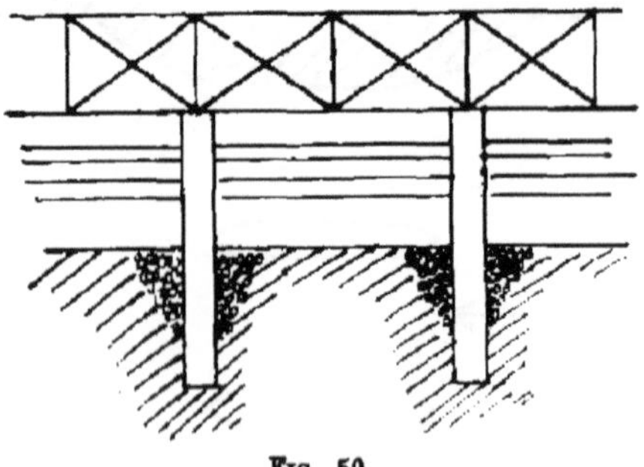

FIG. 50.

depth of the lowest point, found as just explained, will take place in one-third of the span of the bridge. The span can then be so fixed as to give no heading up. It is not assumed that there will be no increase in velocity through the bridge. The velocity in the deep scoured portions will be increased. The piers are protected by loose stone (fig. 50). The spans vary from 100 to 250 feet. The bridge over the river Chenab at Wazirabad had originally sixty-four spans of 145 feet each. The number of spans has since been reduced to twenty-eight. With a very long bridge, the current of the shifting stream is more likely to strike the bridge obliquely, though this is not the chief reason for reducing the length. Long spans, say 250 feet, have been found to be better than shorter spans; the cost of the stone protection round the piers is of course less (*Government of India Technical Paper*, No. 153, "River Training and Control on the Guide Bank System," by Sir F. J. E. Spring, C.I.E., 1904).

2. **Syphons and Culverts.**—Syphons are used to pass drainage channels or other streams under canals or other lines of communication. In the case of a masonry syphon under a stream which may be dry while the syphon is full, the weight of the arch and its solid load must be not less than the upward pressure of

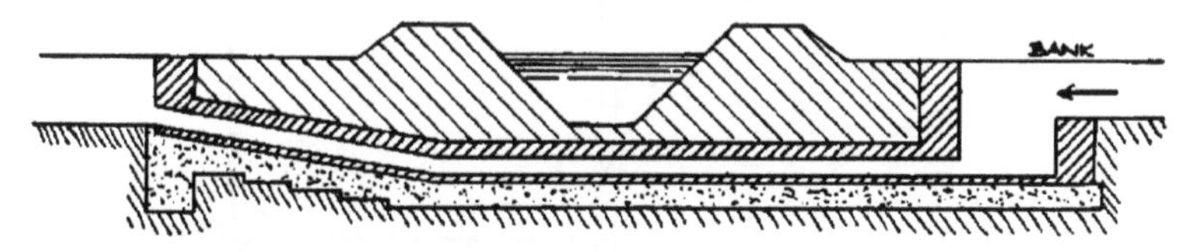

Fig. 51.

the water passing through the syphon. The channel sometimes has a vertical drop at the upstream side (fig. 51) and a slope at the downstream side. The slope enables any solid materials to be carried through, and facilitates cleaning out and unwatering. The drop at the upstream side does not give rise to any shock on the floor when the syphon is full, but a slope is preferable if there is room for it, and it causes less loss of head.

A culvert which is liable to run full and which has a steep approach channel (fig. 52) may become suddenly drowned on the upstream side. As soon as the water rises to the crown of the arch, the wet border of the culvert increases and this reduces the velocity and discharge. The water coming down the approach channel then rises abruptly, and the increased section of the stream causes a reduced velocity of approach, and this further reduces the discharge through the culvert. The heading up continues until the difference in the up-stream and downstream water-levels is great enough to readjust matters (*Min. Proc. Inst. C.E.*, vol. clxxxvi.). The possibility of this heading up occurring should be attended to in the design. In the case of a culvert in a railway embankment where heavy floods have to be passed, the culvert may be made bell-

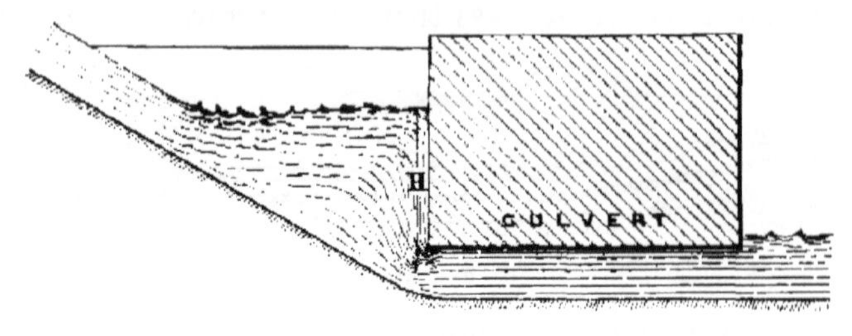

Fig. 52.

mouthed by a curved embankment constructed on its upstream side.

3. **Training Works.**—The object of the upstream and downstream protections already described (Chap. X.) is to prevent damage to the structure owing to the disturbance caused by the structure itself. When a river is given to shifting its course (Chap. IV., *Art. 9*) and cutting away its banks, protection of another kind is required. The stream, if left to itself, may cut away one bank upstream of the structure for a long distance, and eventually damage, or destroy by undermining, the upstream pitching and the abutment itself. This is known as outflanking. If in the neighbourhood of the

line A B (fig. 53) there is nothing for the river to damage,—if, for instance, the structure is a weir with a canal, if any, only on the opposite bank of the river,—and if the land is of no particular value, the case could conceivably be met by protecting the abutment on all sides, but even then there might be a chance of the erosion of the bank continuing until the stream had formed a connection at C with the downstream reach. This, of course, in the case of a weir, would render the work useless and might even destroy it.

In the case of a bridge carrying a road or railway, or of a syphon or aqueduct carrying a canal or other stream, it is wholly inadmissible to allow the stream to cut away even as far as the point A for fear of its severing the line of communication. Thus in every case it is practically necessary to prevent any serious erosion of the bank upstream of the structure. In ordinary cases it is sufficient to protect the bank C D by any of the methods given in CHAP. VI., *Art. 3*, the protection being turned inwards, as shown at D, to prevent the end of it being damaged.

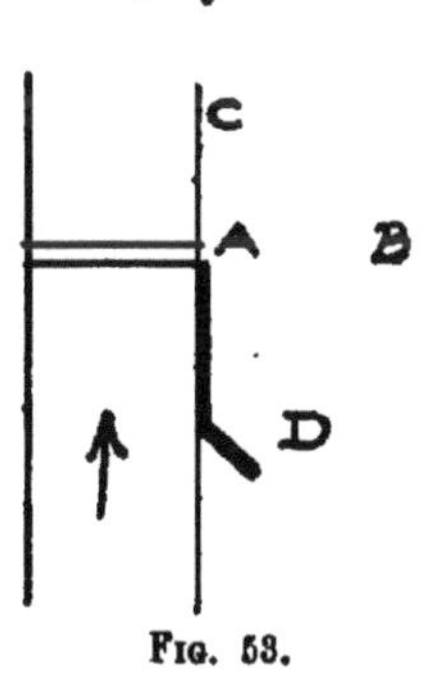

FIG. 53.

In the case of railway bridges across the great shifting rivers of India, protection used at one time to be afforded by various systems of spurs. This has now been abandoned in favour of Bell's guide banks (fig. 54), which are found to be far more satisfactory. These guide banks are discussed in the paper by Spring quoted above (*Art. 1*). The spaces behind the guide banks become filled with water, at least during floods, and are meant to be silted up. An opening in the railway embankment should be provided at A, and another on the opposite side of the river, to ensure a constant flow of water (CHAP. V., *Art.* 3), but they should not be large enough to cause high velocity. The chief danger to which a guide bank is liable is outflanking when the stream assumes the position shown. To guard against this danger it is necessary

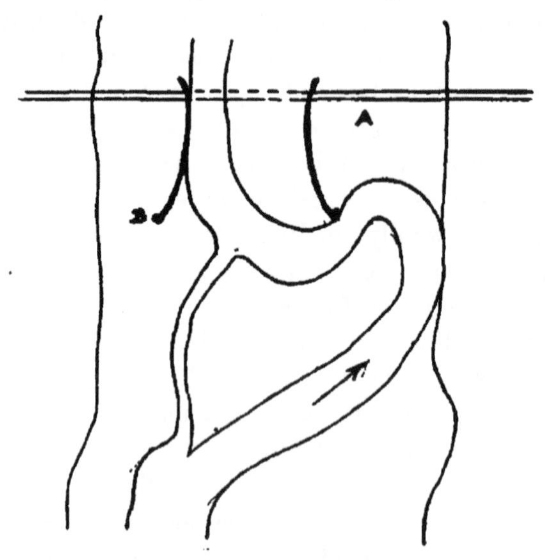

FIG. 54.

to have very strong and massive heads to the guide banks. When the bank of the eroding stream, downstream of the guide bank head, becomes a semicircle or thereabouts, the stream takes a short-cut across the sandbank, and to encourage this an artificial cut can be dug, at the season of low water, on any suitable line.

If the guide banks were made with an increased width of opening at the upper end, this would reduce the chance of outflanking but would increase the danger from a direct attack such as indicated, in the figure, on the left bank. It has been suggested that the width at the upstream end should be less than at the bridge, but this seems undesirable. Probably the form shown in fig. 54 is the proper one. The length of the guide bank upstream of the bridge is made about equal to the span of the bridge between the two guide banks. If made less than this, the river might cut into the line of

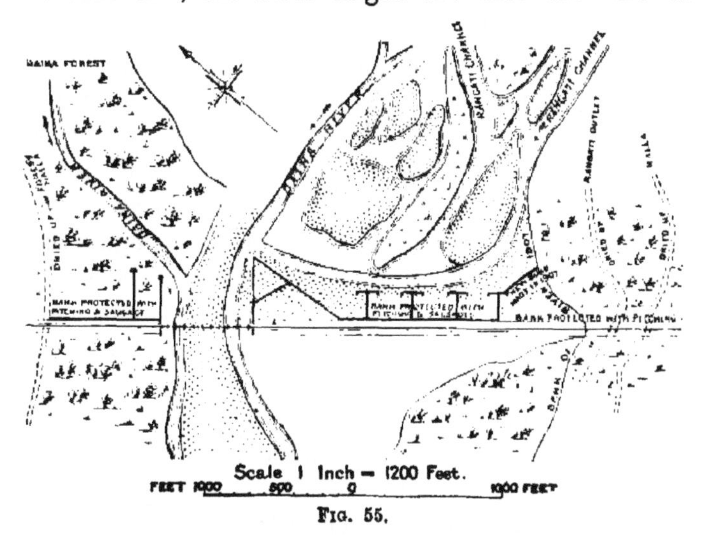

Fig. 55.

railway. The length of guide bank downstream of the bridge is generally 300 to 500 feet, being greater as the velocity of the river is greater and the sand of its bed finer.

The Bengal Dooars Railway runs near the foot of the Bhutan Himalayas, and crosses some broad river channels which, after the excessively heavy rains which occur, are filled by streams of very high velocities. One such channel or set of channels (fig. 55), more than half a mile wide, is provided with a bridge whose waterway consists of ten spans of 60 feet each. The railway embankment across the remainder of the channel having been breached in many places in 1903, protection was afforded by T-headed spurs and other groynes, the first arrangement, which withstood the floods of 1904, being as shown in the figure. The triangular apex of the A-shaped groyne, south-east of the bridge, was added in 1905 because, in its absence, the water struck the bridge obliquely. After the addition there was a great deposit of silt in the neighbourhood of the four T-headed spurs. Next year the river, in a great flood, rose over the top of the railway embankment near these spurs, and finally caused a breach 600 feet wide. The embankment was afterwards raised. The velocity through the bridge seems to have approached 18 feet per second. The bridge had at first no floor. A floor was added, but was much damaged by the floods (*Min. Proc. Inst. C.E.*, vol. clxxiii.). The level of the floor is not given, but it would seem to have been desirable to make it at a very low level. The rising of the stream over the railway embankment was attributed to the silting up near the T-headed spurs. The addition of the triangular portion above referred to would seem to have somewhat assisted this process. If all the trouble could have been foreseen, it might have been best to build an additional bridge 2000 feet south-east of the existing bridge. The groynes were composed of the wire-network rolls, described in CHAP. VI., *Art. 3*, piled pyramid fashion.

CHAPTER XII

DRAINAGE AND FLOODS

1. **Preliminary Remarks.**—*Arts. 2* and *3* of this Chapter deal with the calculation of flood discharges, *Art. 2* dealing with small streams, in which the water has to be got rid of, and *Art. 3* with large streams. The remaining articles discuss the methods of predicting floods and of preventing them from doing damage. When the discharge figures have been arrived at in any case, the necessary masonry works can be designed in accordance with the principles described in CHAPS. X. and XI. For remarks regarding the design of channels and banks, see CHAP. IX., *Art. 4*, and also *Art. 6* of the present Chapter.

In England, land near a stream or flooded area is said to be "awash" when the flood water rises to within 3 feet of the surface of the ground. The drainage of such land is apt to be unsatisfactory. If land is flooded or awash, it may be desirable to shift the outfall of a branch drain to a point lower down in the main outfall.

2. **Small Streams.**—In dealing with small streams, such as branch drains or natural streams not far from their sources, the engineer is concerned only with their maximum discharges. He has to design culverts,

bridges or syphons to pass the streams under roads or other works, or to design channels or waste weirs for them. In a settled country there may be already some works in existence on the same stream, and these may form a guide, or it may be possible to obtain local information as to the height or volume of floods. Even in such a case rainfall figures will be most useful. In districts where there is no settled population, and in any case where the stream is ill-defined, and the flow fitful, the rainfall figures may afford the only, or at least far the best, means of estimating the discharge.

The rainfall to be considered in all these cases is the maximum likely to fall in a short period of time. The catchment areas dealt with are small, say 5 square miles or less. It must be assumed that the fall of rain extends to all parts of the catchment area, and that its duration is sufficient for the water from all parts of it to reach the site of the work. The different valleys or divisions of the catchment area should be considered separately, and regard must be had, not only to the area of each division, but to its length and declivity measured along the course of the stream which drains it. On these two factors depend the time taken by the rain water to reach the site of the work. The rate at which the rain water flows over the ground into rills or small subsidiary streams may be taken to be ¼ mile per hour in flat land, and 1 mile per hour on steep hill sides. The velocity of the current in the rills and larger streams is generally 2 to 4 miles per hour. It can, when necessary, be calculated roughly from the size and slope of the stream. To be on the safe side, the highest probable figure can be taken.

The time taken by the water to flow from the

furthest points of the catchment area to the site of the work having been arrived at as above, the next thing is to estimate the probable maximum intensity of the rainfall during that time over the whole catchment area. The only figures immediately available will be the mean annual rainfall, or perhaps the maximum fall in twenty-four hours, but it has been shown (CHAP. II., *Art. 5*) how the probable maximum fall over a shorter period may be estimated.

The next thing to be calculated is the "run-off," *i.e.* the probable proportion of the rainfall which will at once run off. This may be less than the proportion which will eventually become "available," because some of it may go to feed the underground supply from which springs are fed. The proportion running off a small area in a short time would, under most circumstances, be rather difficult to estimate, but in the case under consideration, only the probable maximum figure is required. This occurs when the ground is saturated. Under these circumstances the ratio of the run-off to the total fall may be somewhat as follows:—

Steep rocky hillsides	·70 to ·90
Ordinary hills	·50 „ ·75
Undulating country	·40 „ ·50
Flat country	·30 „ ·35

The figures can be increased when the surface is specially hard or frozen, and decreased when it is soft, sandy, covered with woods or vegetation, or cultivated.

Whether or not the above procedure is necessary in its entirety depends chiefly on the size of the proposed work and on the degree of inconvenience likely to arise from any wrong estimation of the discharge.

In designing syphons to carry torrents across the Upper Jhelum Canal in the Punjab, the discharge from a catchment area of ·79 square miles was found to be about 4000 cubic feet per second. This is at the rate of about 5000 cubic feet per second per square mile, and is equivalent to a run-off of 7·8 inches in an hour. The catchment area was among low hills, not far from the Himalayas, and the declivities of the rills were very steep. The superintending engineer, Mr R. E. Purves, states[1] that the discharge observations were reliable, and that falls of rain of an inch in ten minutes occur not infrequently, even though the fall in twenty-four hours might not exceed 2 or 3 inches. In order to account for the discharge in the case under consideration, it would at first seem to be necessary to assume not only that a fall at the rate of 7·8 inches per hour had occurred, but that the whole of it had run off. It is not, however, necessary to assume quite so much. The ground being saturated, the rain falling in a period of five minutes might be reaching the discharge site with little loss. A suddenly increased fall at the rate of 6 inches per hour might then occur, and the water travelling more quickly and with hardly any loss, would overtake that already passing the site. This case seems to show that for a very small catchment area the whole of the fall, and more, must be allowed for.

The Chief Engineer of the Punjab did not accept the above figures.[2] He remarked that observations taken under great difficulties as to time and place are liable to error, and he considered that an allow-

[1] *Report on the Revised Estimate, Upper Jhelum Canal.*

[2] *Revised Estimate of the Upper Jhelum, Upper Chenab, and Lower Bari Doab Canals.*

ance of rainfall at the rate of 4·8 inches per hour —a rate which had been observed elsewhere—and a run-off of ·75 of the fall would be sufficient. He accepted a discharge of 2000 cubic feet per second for catchment areas of less than 5 square miles, assuming the run-off to be ·75 of the fall, but afterwards increased the figure to 2400 cubic feet per second. The chief engineer did not overlook the fact that in the designs for the drainage aqueducts a free-board of 5 feet had been allowed, and perhaps this led to an acceptance of an estimated discharge less than would otherwise have been accepted. It does not seem to be at all certain that the figure put forward by Mr Purves was far wrong. When the original project estimate for the Upper Jhelum Canal was framed, the irrigation engineers had had no experience of small and steep catchments, and no one had suspected that the discharge per square mile would be anything like the above. The sums of money provided for works for the passages of torrents had to be increased in ratios varying from 2·5 to 1 to 6 to 1.

The following statement shows the figures for other small catchment areas in the neighbourhood of the Upper Jhelum Canal:—

Catchment Area.	Discharge per square mile.	Run-off.
Sq. miles.	Cub. ft. per sec.	Inches.
·79	5000	7·8
1·47	3825	5·82
2·96	2214	3·46

In the south-east of New South Wales flood discharges of 135 and 84 cubic feet per second have been found for catchment areas of ·91 and 2·5 square miles respectively in broken country.

3. **Rivers.**—It is possible to apply the methods of the preceding article to large catchment areas, but the results would be quite unreliable. If the calculations were made so as to err on the side of safety, the resulting discharges would often be enormous. The following table shows some figures based on actual flood discharges. None of the localities have excessive rainfalls, though most are liable to occasional very heavy falls. In mountainous districts in the North of England and in Scotland the flood discharges per square mile of catchment area have been found to vary from 64 to 320 cubic feet per second.

Reference Number.	Country.	Locality.	Catchment Area.	Flood Discharge per sq. mile of Catchment Area.	Remarks.
			Sq. miles.	Cub. ft. per sec.	
1	India.	Upper Jhelum.	5 to 10	1613	
2	"	Nagpur.	6·6	480	
3	South Africa.	Near Cape Town.	34·5	78	
4	"	Near Port Elizabeth.	35	640	Estimated.
5	New South Wales.	South-East District.	49	37	
6	India.	Upper Jhelum.	56	1000	
7	"	"	174	550	
8	New South Wales.	South-East District.	418	11·2	
9	India.	Kali Nadi Stream.	2593	51 or more	Estimated roughly.

The tendency of the figures in column 5 of the table is to decrease as the catchment area increases. This tendency has long been known, and attempts have been made to found on it formulæ for calculating flood discharges. One such formula is $Q = c\ M^{\frac{3}{4}}$ where Q is the flood discharge in cubic feet per second and M is the area of the catchment in square miles. The formula is roughly correct, *c* being a constant for catchment areas of not dissimilar characters and with rainfalls not differing much. But for other cases there is no knowing how *c* may vary, and this renders the formula practically useless. The author of another such formula quotes cases Nos. 5 and 8 in the above table, and the two cases mentioned at the end of *Art. 2* as agreeing fairly well with the result of his formula. The tendency just mentioned is due to the fact that every river is composed of tributaries which have their own small catchment areas but are, when measured to the general outlet or point where the discharge is under consideration, of very different lengths, to the improbability of heavy rainfall occurring over all these small areas at such times as to cause the different flood waves to arrive simultaneously at the outlet, and to the facts that in the case of the longer tributaries the flood waves flatten out (*Hydraulics*, Chap. IX., *Arts. 3* and *4*) so as to arrive more gradually, and that, unless rain is also falling all along their courses, these longer tributaries undergo losses from evaporation and absorption. But occasionally it happens that the various flood waves do arrive at the outlet more or less simultaneously, and that the rainfall continues so long and is so widely distributed—though not necessarily of the same intensity as that which caused the flood—that the flood waves do not flatten out and that losses in the channels do not occur. Floods can thus vary to an extraordinary degree in severity, and formulæ are quite useless. This is why floods occur surpassing all previous records, as, for instance, the recent floods in Paris. However severe a flood may be, it can never be said that the maximum has, even probably, been attained unless it can be shown that the rainfall has been so heavy, so long continued, and so distributed that anything worse is not likely to occur.

The best method of estimating the flood discharge of a large perennial stream is to ascertain, by local inquiry, the height to which it is known to have risen, and to take cross-sections of the channel and calculate the discharge (CHAP. III., *Arts. 4* and *5*). In designing works, allowance can be made for a flood exceeding any known before. This method applies also to a case in which a river is formed by the junction of two or more large tributaries. It is possible that the tributaries have not, within the memory of man, been in high flood simultaneously. If so, the chances of this occurring are no greater and no less than if the stream was composed merely of a number of small affluents. Remarks regarding intermittent streams are given in CHAP. III., *Art. 7*.

Since an acre contains 43,560 square feet, and a twelfth of this is 3630, it follows that a fall of 4 inches of rain, of which 1 inch runs off, in an hour, gives a discharge of 3630 cubic feet per hour, or about 1 cubic foot per second. This is 640 cubic feet per second for a square mile. The figures in column 5 of the above table show that the run-off was, in the cases quoted, generally far less than 1 inch. In case No. 4 it was 1 inch, and in case No. 2 it was $\frac{3}{4}$ inch.

In the case of the Kali Nadi (No. 9 in the table) an aqueduct to carry the Lower Ganges Canal over the stream was being designed. The flood discharge, estimated from the supposed flood-level and cross-section of the stream was (*Min. Proc. Inst. C.E.*, vol. xcv.) 26,352 cubic feet per second. The discharge, estimated by assuming a fall of 6 inches of rain in twenty-four hours over the catchment area—then believed to be 3025 square miles—and a run-off of ·25 of the fall, was 114,950 cubic feet per second. This figure was rejected on the ground that the rainfall would not be continuous over so large an area as 3025 square miles. An allowance of 7 cubic feet per second per square mile was made and, a fresh survey having shown that the catchment area was only 2593 square miles, a discharge of 18,000 cubic feet per second was allowed for. The aqueduct was built, about the year 1875, with five arched spans of 35 feet each, the total area of the waterway being about 3000 square feet. The length of the piers and abutments was 212 feet, the width of the canal carried over the aqueduct being 192 feet. In 1884 the aqueduct was partly destroyed by a flood whose discharge was about 44,000 cubic feet per second. In July 1885 it was wholly destroyed by a flood whose discharge was estimated at 132,475 cubic feet per second, but was probably more. The discharge must have been more than 51 cubic feet per second per square mile. The aqueduct was rebuilt with a waterway of about 15,000 square feet. Below the aqueduct there was a bridge which had been standing for a hundred years. Its waterway was only 1146 square feet. It was not much damaged by the flood of 1884, but much of the water passed round it, breaking through the embanked roadway or pouring over it. It is understood that the bridge was destroyed by the flood of 1885.

This case shows the necessity for making every possible allowance in calculating flood discharges for important works. The smallness of the discharge, as calculated from the cross-section of the stream, was probably owing to its being dry when the survey was made, so that the velocity could not be observed, but it is probable that such a discharge as wrecked the aqueduct had never before passed down the stream.

4. **Prediction of Floods.**—At any place high up on the course of a stream, the occurrence of a flood can often be predicted when rain storms—often accompanied in the tropics by lightning—can be seen to be occurring towards the sources of the stream. For any station lower down the stream and for precise information in any case, the readings of gauges higher up the stream can be telegraphed. If the station is at a great distance from the gauge and if there is railway communication, the readings can be sent by post.

In order to be able to predict the time of the arrival of a flood at the lower station the reading of a gauge there, and also of that at the upper station, should be taken at frequent intervals. In the case of large rivers and distances of hundreds of miles, the interval may be six or even twelve hours, but in other cases it should be much less. If the readings are plotted, as in fig. 56, oblique lines can be drawn to connect the saliences and depressions, and the time taken by each change can thus be readily seen. When the upper part of the stream is formed by two or more important tributaries there should be a gauge in each.

As to what constitutes a flood, the gauge diagram of a river (fig. 56) is generally such that a line can be sketched as shown dotted. The rises above this line are floods. The maximum flood discharge of a Northern Indian river is estimated roughly as being 100 times the low-water discharge. Leslie's rule for floods in the British Isles is that if all the daily discharges of a stream during the year are ranged in order of magnitude, the discharges of the upper quarter are considered to be floods.

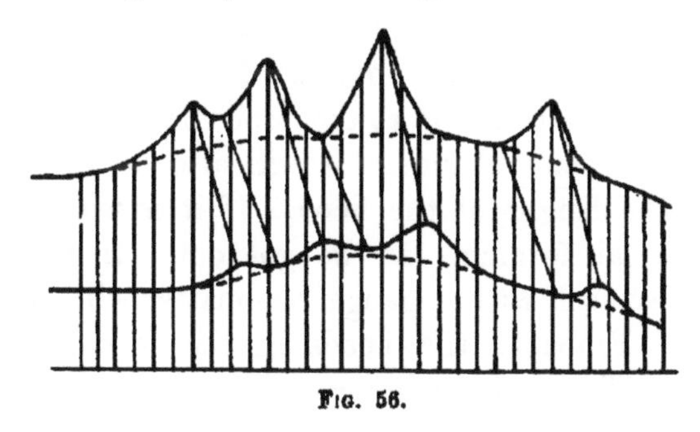

Fig. 56.

In India it is sometimes arranged that a telegram shall, in the low-water stage of the river, be sent from the upper station when a rise of 2 feet occurs in twenty-four hours or any less period, with a further telegram for any such subsequent rise. The telegram states the exact reading on the gauge and whether the water is rising steady or falling. This is given as indicating the procedure that may be followed where the telegraph has to be used, but when long and frequent telegrams are not desirable.

The advancing end of a flood wave may, while the

wave is rising and being formed, travel rapidly, but when the wave has been formed it travels at the ordinary rate of flow of the risen stream. The advancing end of a trough may, while it is being formed, travel rapidly, but after formation it travels at the ordinary rate of the fallen stream (*Hydraulics*, CHAP. IX., *Arts. 3* and *4*). Thus the rate at which a change in water-level travels down a stream depends at first on the amount of the rise or fall, but afterwards on the water-level of the risen or fallen stream.

By taking the above facts into consideration and noting the actual times obtained from the diagram, it will be possible to arrive at the probable time that will be taken by any change. It will also be possible to predict the height of the flood. If it is worth while, an empirical formula can be got out. If there are tributaries, each with a gauge, the matter will be more difficult. Probably the floods in the tributaries will arrive at different times, but even in such cases empirical formulæ have been arrived at, especially in France, and are mentioned in various volumes of the Proceedings of the Institution of Civil Engineers.

In all cases predictions are liable to be more or less upset if rain falls in the tract between the upper and lower gauges. In very dry weather the speed of a flood wave may be somewhat reduced, and the height to which it rises will almost certainly be reduced.

The full effect of a change will not be felt at the lower station unless the change at the upper station is maintained for a sufficiently long period. A short wave or trough flattens out. Thus in any empirical formula or system of prediction, the time over which the change extends at the upper gauge must be taken into account, or else there must be several upper gauges and the readings of all of them be taken into account.

In mountainous districts landslips sometimes occur and block the valley of a stream which then forms a lake. The water gradually rises and eventually flows over the dam and sweeps it away causing a flood, which is of great suddenness and height but decreases very quickly in height as it travels down the valley. In a case which occurred in the Himalayas in 1888 the inhabitants of the valleys, from the dam to the point where the river debouches from the hills, were compelled by Government to vacate all habitations below the probable level of the flood, and no loss of life occurred. Similar floods, but on a smaller scale, may be caused by the bursting of ordinary reservoir dams. In some continental rivers ice may obstruct the stream and cause floods.

5. **Prevention of Floods.**—The extended use of field drains has, in recent years, done much to increase the severity of floods in England and other countries. One method of mitigating or preventing floods is the construction of reservoirs for storing the water. Reservoirs locally known as "washes," formed by setting back the embankments, exist on the Fen rivers. One wash, on the Nene, below Peterborough, is 12 miles long and half a mile wide and is filled, in floods, to a depth of 7 feet and holds 1 inch of rainfall over the river basin, and this is found to be sufficient. Reservoir construction is, however, in most cases, impracticable owing to the expense. To store the water which is given by 1 inch of rain in the basin of the Thames, a reservoir would be needed 50 feet deep and covering about 7 square miles. It might cost £7,000,000.

The afforestation or reforestation of river basins (CHAP. II., *Art. 4*) is also occasionally undertaken, but is not generally practicable.[1]

The most practicable methods for preventing flooding are lowering the water-level of the stream and constructing embankments along it. These will be considered in the next two articles.

6. **Lowering the Water-Level.**—The water-level of a given length of stream can be lowered by lowering the bed, widening the channel or straightening the channel. The efficiency of these processes is in the order named. As stated in CHAP. I., *Art. 4*, the alteration to the channel must in any case be continued to some point downstream of the reach under consideration. Let the channel be supposed to be of "shallow" section with sloping sides. Let W be the mean width, D the depth, and S the slope. Let it be required to lower the water-level by an amount equal to $\frac{D}{5}$. This can be effected by lowering the bed by about 25 per cent. of D, or by increasing the width by about 50 per cent., or by increasing the slope by about 100 per cent. If the bed is lowered, V is not affected, and the mean width is reduced. Increase in W reduces D, and therefore reduces the hydraulic radius and the velocity. Hence the large amount of widening necessary. When S is increased the velocity, if R remains the same, is affected only as $\sqrt{S}$ (*Hydraulics*, CHAP. VI., *Art. 2*), but the depth of water is reduced and R therefore reduced. Dressing the sides of a channel, so as to make it smoother, produces the same effect as a slight widening.

It does not, of course, follow that lowering the bed

[1] See also Note on p. 161.

is always the best plan and straightening the worst. Any one of the processes may be more or less impracticable because, for instance, of the hardness of the material to be removed, or the expense—including compensation—of removing obstructions.

A particular kind of widening consists in digging a new channel and keeping both the new and the old channel open.

If a channel contains a weir, or a local raised portion of bed forming a kind of submerged weir, or a contracted place or narrow bridge, the upstream water-level can be lowered by simply removing or reducing the obstruction. The lowering of the water-level will be greatest at the site of the obstruction, and will be zero at some point far upstream (*Hydraulics*, CHAP. VII., *Art. 5*). If the raised portion forms a long shoal, its removal—supposing its height above the general bed to be the same—will have more effect than if it were short. If the height of the raised portion is small compared to the depth of water, or the amount of contraction small compared to the width of the stream, the removal may have much less effect than might appear (CHAP. I., *Art. 4*).

In soft soils one advantage of the straightening system for lowering the water-level is that short-cuts can be dug to a small section, and left to enlarge themselves (CHAP. VII., *Art. 1*).

Another advantage is that after any diversions have enlarged themselves to the size of the rest of the channel—or have originally been so excavated—the whole channel may scour, and the water-level continue to fall. This, of course, should be allowed for if likely to occur.

The same thing may occur in the case of the removal of a weir, shoal, or contracted piece of channel. The scour will act at first close to the site of the obstruction, but it may work upstream.

In widening or deepening a channel for the purpose of mitigating floods, it is a good plan to begin work at the downstream end, because the lowering of the water-level will extend upstream beyond the reach in which work is done, and this may facilitate work further upstream. As regards any tendency for a widened reach to silt up again, any such silting is not likely to be great in a short period of time, and need not prevent the carrying on of work in various reaches, if this is convenient.

7. **Flood Embankments.**—A flood embankment may be close to the edge of the river or it may be set back. If set back it need not follow all the windings of the stream. The setting back of an embankment gives an increased waterway to the stream during floods, and therefore a lower flood-level, but the effect of this is trifling in cases where the depth of the water on the flooded land is small, especially if such land is covered with vegetation, or is otherwise much obstructed. Setting back is generally necessary in cases where the stream is liable to erode the banks to any considerable extent. In such a case the embankment should not be so near to the river as to be in much danger from erosion, but the ground, as already stated (CHAP. IV., *Art. 9*), generally falls, in going away from the river, so that when an embankment is set well back it is in lower ground, more expensive and more liable to breach. The most suitable alignment is a matter of judgment, and depends largely on the extent to which the river is likely to shift.

Embankments should, where possible, be made in straight or properly curved reaches. A flood embankment, at least at its upstream end, should terminate in ground which is above flood-level. The top of an embankment should be, in the case of a large river, 2 or 3 feet above the high flood-level of the river. It should, of course, be graded parallel to the general high flood-level, but neither the gradient nor the height of the flood is usually known with accuracy (CHAP. II., *Arts. 1* and *2*). There is generally a record or mark of some high flood, and this is taken provisionally as the flood-level. Or the level is calculated approximately from the flood readings on the nearest river gauge. If experience shows that the embankment is too low, it is raised. The cross-section of an embankment depends on the soil, on the extent of damage which results if a breach occurs, on the funds available, and on the value of the land which the embankment has to occupy.

Where an affluent enters the river it will probably be necessary to run out branch embankments. Sometimes cross embankments are run from the main embankment to high land. Their object is to localise the damage if a breach occurs. Along the back of the embankment there may be a drain and it can be made to discharge its water, when the river is not in flood, through the embankment by means of sluices or by pipes closed by flap valves which will not allow flood water from the river to pass through. There may be sluices in the embankment for the purpose of irrigating the land at the back.

The immediate effect of the construction of flood embankments along a river is to raise the water-level, because the floods can no longer spread out over the country, but this effect will not be great if the sectional area of the flood water was small or its velocity low. The river may or may not tend, after the construction of flood embankments, to raise or lower its bed. It has already been remarked that questions of silting or scouring cannot be answered in a general manner. In the case, however, of floods spilling over a piece of country, the depth of the flood water is generally small and the country more or less obstructed. Some deposit of silt generally occurs. The construction of an embankment reduces the area of the flood water, and thus generally reduces the silting and leaves more silt in the river proper. The depth and velocity in the river are increased. Everything depends on which is increased most. Most likely the stream is of shallow section and the velocity is increased most (CHAP. IV., *Art. 6*, par. 6), and the increased silt-supporting power may make up for the increased charge of silt.

Sometimes when a main embankment is set far back, a subsidiary embankment of smaller section is constructed closer to the stream. This is often objectionable. The smaller embankment is liable to breach, and the water then rises suddenly instead of gradually against the main embankment, which is thus endangered to some extent, especially as it is dry instead of being soaked.

It is often said that one effect of embanking a reach of a river is to increase the severity of floods further downstream. The importance of this is generally exaggerated. The narrowing of the flood stream in the embanked portion causes the flood to travel more quickly and rise higher in that particular reach. At a place further downstream the same effect is produced, but in a less degree and only because of the increased velocity and consequent reduction in the flattening out of the flood wave, especially when the rise is soon succeeded by a fall. When there is a gradual rise lasting for a considerable time—and this is most likely to cause a high flood—there is no rise of the flood-level downstream of the embanked reach, except such as is due to the increase in the discharge of the stream consequent on the absorption and evaporation being less than before, owing to the reduced area of flooding in the embanked reach. In the case of a long-continued rise, such as that just mentioned, it is the reach immediately upstream of the embanked reach which will, to some extent, share in the increased height of the floods.

An embankment may suitably have side slopes of 4 to 1 on the river side and 3 to 1 on the land side, with a top 10 feet wide and 3 feet above high flood-level. On the Irrawaddy the top width is generally 8 feet. For very high and very low embankments it is 10 feet and 3 feet respectively. In Holland 1 foot above high flood-level was at one time supposed to be the rule, but in practice it was usually 4 feet. With sandy soil the riverward slope prescribed was 6 to 1. Such flat slopes are not necessary if fascining or stiff soil is used as a protection. On the Rhine the top width of embankments consisting of gravel and sand has been made about 15 feet, but the side slopes were 1½ to 1 and 1 to 1. The embankments had spurs to keep off the current.

Sand, protected as above, makes a good embankment, and rats do not burrow into it. Of course, if a breach occurs in an embankment consisting mainly of sand, it will enlarge very quickly. In some cases an embankment has a core wall of sand or of clay puddle. In

Holland, on sandy soil, a trench 8 feet wide is made and taken down to the clay.

Embankments require to be made with great care. The earth should be deposited in layers. In Holland, horses are driven up and down over each layer. In some parts of India the earth for embankments is brought from the borrow pits by scoops drawn by bullocks. The earthwork is of so excellent a character, owing to the earth being trodden down, that no settlement has to be allowed for. Where the soil is sand the top and faces of the embankment should be of good stiff soil, if it can be obtained, for a thickness of 9 inches or a foot, or else the face next the river should be protected by fascining (CHAP. VI., *Art.* 3) for 2 feet above, and several feet below high flood-level. Such protection may be necessary in any case where waves are liable to occur. In Holland embankments are turfed, and trees and shrubs are not allowed to grow. In the Punjab the growth of all kinds of jungle is encouraged. It binds the soil together and protects it from the wash of waves and from winds which blow away sand and dust, and so wear the embankment slowly away.

In embanking a long reach of a river it is convenient to begin from the upstream end, because otherwise floods may get behind the finished part of the embankment and, becoming impounded in a "pocket" formed by the embankment and high land, rise to an abnormal height and, unless gaps in the embankment have been left or are subsequently made, cause breaches.

During high floods pegs should be driven in at frequent intervals, to mark the high flood-levels. If a higher flood occurs, the peg is shifted. The levels of the pegs can be observed at leisure.

When a breach occurs in an embankment, the first thing to do is to protect the ends so that the breach shall not lengthen. If the water passing through a breach becomes pocketed, the embankment may have to be cut to let it out.

Regarding the stoppage of leakages, see CHAP. IX., *Art. 1.* Regarding the closure of breaches, see CHAP. VII., *Art. 2.*

For a description of flood embankments along the great shifting rivers of Northern India, see *Punjab Rivers and Works.*

Note to Art. 5.—Floods can sometimes be mitigated by sinking pits in the flooded area so that the flood water comes in contact with permeable strata and is absorbed by them.

CHAPTER XIII

RESERVOIRS AND DAMS

1. **Reservoirs.**—The object of a reservoir is to store water for town supply or for irrigation or other purposes. Reservoirs for the water supply of towns are divided into "impounding reservoirs" and "service reservoirs," the latter being of comparatively small size, and their object being to store, near to the town, a supply sufficient for a short period. Instead of one impounding reservoir there may be several, formed by various dams and one discharging into another. When a reservoir is mentioned without qualification, an impounding reservoir is meant. A reservoir is generally made by blocking up a valley by means of a dam of earth or masonry. The site of the dam should be selected at a place where the valley is narrow. The lowest portion or "bottom water" of a reservoir is usually not drawn upon, because it is less pure than the rest, and it has to be left, in dry weather, for the fish. It is not included in calculating the capacity of the reservoir.

In Great Britain, when the water of a stream is impounded, "compensation water" has to be given back to the stream lower down. This compensation water is generally given in the form of a constant supply, and amounts to perhaps a quarter of the available supply. It has to be included in calculating the daily supply taken out of the reservoir. The advantage to the stream in having this addition to it during dry weather is very great.

It has already been seen (CHAP. IX., *Art. 1*) that in an earthen bank which has to retain water the leakage generally decreases rapidly and the bank becomes almost impermeable. The same is true of the surface of a valley, in the case of most ordinary soils, provided that it is kept submerged. Any portions which become exposed to the sun and weather are likely to crack and give rise to percolation. Thus a reservoir formed by the construction of a dam resting on the surface of the ground may be more or less water-tight according to circumstances. There are many which are sufficiently water-tight. But in most cases the dam—or an impervious core-wall—is carried down to an impervious stratum. A masonry dam is carried down to rock.

In the case of dams of considerable height the soil should be examined by borings. If there is an inclined stratum not well connected with that below it, unequal settlement of the dam may occur; and this may also happen if there is a thick stratum of clay, owing to its compressibility.

Except for very high dams—those, for instance, more than 110 or 120 feet in height measured from the ground to the water-level—an earthen dam is cheaper than a masonry dam. It is also more easily raised and strengthened—though this operation has also been effected on masonry dams—in case, for instance, of the silting up of the reservoir, a process which is slow in

England, but not so slow when water containing much silt is dealt with, as in the case of irrigation reservoirs in India. Sometimes a dam consists of a wall of masonry or concrete with earth behind it as a support. Whatever kind of dam is used, its construction always demands very great care. Serious disasters, with much loss of life, have occurred owing to failures of dams.

A reservoir with an earthen dam is provided with a waste weir for the purpose of passing off flood water, which might otherwise overtop the dam and destroy it. Generally the waste weir is a continuation of the line of the dam. Its crest has to be below the high-water level of the reservoir, but not lower than can be helped, and its length has therefore to be considerable. Sometimes it is provided with grooved piers between which planks are placed in the season when floods are not likely to occur.

In connection with irrigation reservoirs in Western India, it has been pointed out by Strange (*Min. Proc. Inst. C.E.*, vol. cxxxii.) that a long high-level waste weir is best suited to cases in which the replenishment of the reservoir is uncertain, and that in cases where it is nearly certain, the high-level weir prevents the water-level in the reservoir being quickly lowered in the case of an accident or for the purpose of effecting repairs, impounds the earliest floods, which are most charged with silt, and causes the water area to be a maximum, and therefore gives all floods the maximum time in which to deposit silt. He accordingly suggests that the crests of waste weirs in these reservoirs should be shortened and lowered and provided with falling shutters (this had been done in one reservoir and has since been done in another), and that sluices be added with sills at a still lower level than the lowered crest. These proposals seem to be entirely reasonable, though of course it would be necessary to have skilled supervision over the working of the sluices. Sometimes the waste weir is made in a separate place, being separated from the dam by a saddle.

A masonry dam may act as its own waste weir, the flood water flowing over the crest and down the rear slope; but in cases where heavy floods are liable to occur it is usual to provide a separate waste weir by cutting away the side of the gorge either close to the dam or at some other place.

While a dam is in course of construction arrangements must be made to deal with flood water. Generally the construction of some part of the dam has to be deferred to let the water pass. In the case of a masonry dam it does not much matter what part is thus deferred provided the usual procedure of stepping the work back is followed. In the case of an earthen dam it is best to defer a portion, not in the lowest ground where the dam is highest, but to one side of it, thus allowing the highest part of the dam to be brought up continuously. Temporary embankments and weirs can be constructed to cause the water to traverse the desired route without doing damage. Stepping of the earthwork should be avoided as far as possible. If it has to be adopted, the steps should be small. Sometimes the flood water is conveyed away by means of a "by-wash," by an entirely different route.

In Indian reservoirs the discharge over the waste weir may at times be great. The waste weir is sometimes in the position shown in fig. 57, *a e* being the weir. In such a case a special hydraulic problem arises. In a case where a stream whose velocity is V issues from a reservoir or takes off at right angles from a larger stream there is (*Hydraulics*, Chap. II., *Arts. 19* and *20*) a fall in the water of about $\frac{V^2}{2g}$. The same thing occurs downstream of a weir, at least when there is a clear fall which is vertical or nearly so, so that the water after falling has no horizontal velocity. The water has to be started afresh on its course. In the case represented by the figure, the width of the channel is often restricted because of high ground beyond *f*, and the velocity in

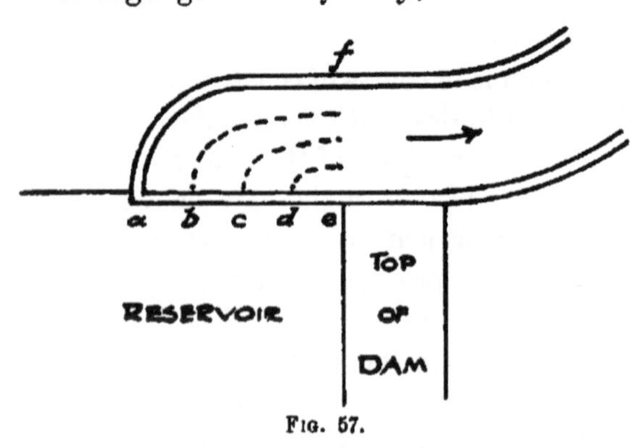

Fig. 57.

the channel may be very high. Suppose the channel below *ef* to be of brickwork with vertical sides, and to have a 20-foot bed, a slope of 1 in 500, and a depth of water of 10 feet. The velocity may be 15 feet per second, and $\frac{V^2}{2g}$ is 3·49 feet. If the water has a clear fall over the weir at *e*, allowance must be made for a depth of water of 13·49 feet, not 10 feet, in the channel at *e*. Ordinarily the length *a e* will be much greater, relatively to *e f*, than shown in the figure. Suppose that *a e* is 300 feet and that the slope of the floor of the channel is carried on at 1 in 500 from *e f* up to *a*, *b*, *c*, and *d*, following in each case the lines marked on the figure which represent the directions of flow. The length *f a* will be about 310 feet, and the floor level at *a* will be about ·62 feet higher than at *e f*. The water-levels below the weir will be in each case 13·49 feet above the floor. This should be allowed for in the design. It is true that the stream on first starting into horizontal motion below the weir moves more or less at right angles to it, and has thus a large sectional area and a low velocity; but it very soon has to turn parallel to the weir and acquire the full velocity of 15 feet per second, and there must be the requisite extra head to give this velocity. If the weir is drowned, the water on passing over it may have a high horizontal velocity, but it will be at right angles to the axis of the channel, and its effect will be wasted in eddies.

2. **Capacity of Reservoirs.**—A reservoir depends for its supply on the yield of a particular valley or valleys which form its catchment area, and the capacity of the reservoir or reservoirs can be altered by altering the height or number of the dams. The need for a reservoir is entirely owing to the inequality in the distribution of the rainfall. If the rain fell in equal quantities week by week, the daily fluctuations could probably be equalised by the service reservoirs. The impounding reservoir could be quite small. Actually, a reservoir is needed to "equalise" the flow—that is, to give a steady flow for an intermittent one. The smaller the reservoir, the sooner it will go dry in a drought and the sooner it will overflow in wet weather and cause waste of the water. In other words, the larger the reservoir the better it will fulfil its function of equalising the flow and the greater the degree to which the catchment area will be utilised.

In the British Isles the distribution of the rainfall which is most trying for a reservoir, occurs when the rain is heavy during the winter and very light in summer. Fig. 58 shows a diagram for a reservoir in the driest year, when the rainfall is (CHAP. II., *Art.* 1) ·63 of the mean annual fall. The distribution of the fall is supposed to be unfavourable as just described. The lower part of the figure shows the water-level at the end of each month, the reservoir being supposed to have vertical sides so that the quantity of water in it is

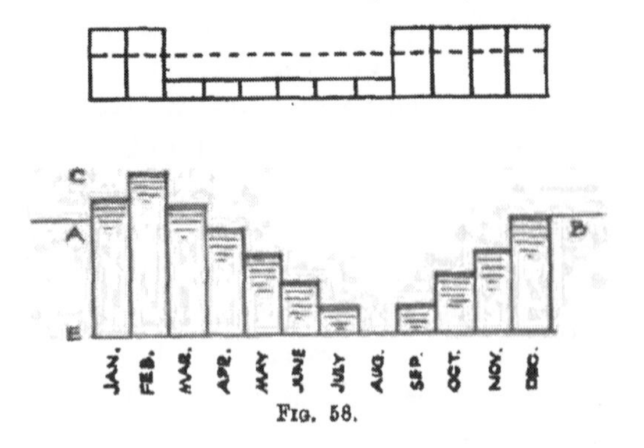

FIG. 58.

proportional to the depth of water. The upper part of the figure shows the water impounded (available fall multiplied by area of catchment) in full lines, and the consumption in a dotted line. The distance between the two lines in any month is the same as the rise or fall of the reservoir in that month. There is supposed to be no overflow, and the total consumption of water in the year is equal to the quantity impounded in the year, so that the levels of the reservoir water surface on 1st January and 31st December, as shown by the horizontal lines A, B at the left and right of the figure, are the same. Deacon, who has investigated the subject, has found (*Ency. Brit.*, Tenth Edition, vol. 33, "Water Supply") that, in order to satisfy the above conditions, the capacity of the reservoir must be 30 per cent. of the water impounded during the year, or about 110 days' consumption. On 1st January the reservoir must be about two-thirds full. At the end of February it is ready to overflow. At the end of August it is just becoming dry. The daily consumption is supposed to be steady throughout the year.

As an instance, suppose the catchment area to be 1000 acres, the mean annual fall 60 inches, with a loss from evaporation and absorption of 14 inches. The available rainfall of the year is (see last column of table below) 23·8 inches, or $1{\cdot}98\dot{3}$ feet. The water impounded and consumed during the year is $1000 \times 43{,}560 \times 1{\cdot}98\dot{3} \times 6{\cdot}25 = 539{,}962{,}000$ gallons. The reservoir capacity must be $\frac{3}{10}$ths of this, or 161,988,600 gallons. This is represented by the height C E. If the mean available rainfall in January and February is 6·3 inches, or ·525 feet, the water impounded during those months is $1000 \times 43{,}560 \times {\cdot}525 \times 6{\cdot}25 = 143{,}931{,}000$ gallons, and the consumption is $\frac{539{,}962{,}000}{6} = 89{,}993{,}667$ gallons. The difference, 53,937,333 gallons, represents the addition A C, to the reservoir. Similarly, the light summer rainfall causes the depletion A E, and the heavy rainfall in the last four months of the year the addition E B. If the height of the reservoir above A B were less than A C, there would be overflow at the end of February; and if the depth below A B were less than A E, the reservoir would go dry before the drought ended. If the capacity of the reservoir were increased either at the top or bottom, the cost would be increased and nothing would be gained. It is not meant that the highest and lowest levels of any reservoir designed as above would always, in the driest year, exactly correspond with the points of overflow and going dry, but they would do so nearly. Deacon states that such a reservoir would fail only once in fifty years, and then only for a short time.

The reservoir considered above does not, as already remarked, fully utilise the yield of the catchment area. In a wetter year there would be overflow and the yield from the reservoir would not be much increased. In order to equalise the flow of the two driest years the capacity of the reservoir must be increased, its yield being also increased, and so on for larger groups of years. By collecting information for large numbers of places in the British Isles, Deacon has prepared diagrams and tables which show the capacities and yields of reservoirs. The following table gives the figures for the case where the rainfall is 60 inches and the loss by evaporation and absorption 14 inches :—

Number of Driest Consecutive Years the Flow of which is to be Equalised.	Net Capacity of Reservoir for a Catchment Area of 1000 acres.	Daily Yield of Reservoir.	Column 2 ÷ Column 3 or Number of Days' Supply contained in the Reservoir.	Ratio of Rainfall to Mean Annual Fall.	Available Rainfall.
	Gallons.	Gallons.			Inches.
1	166,000,000	1,475,000	113	·63	23·8
2	258,000,000	1,815,000	142	·72	29·2
3	329,000,000	1,987,000	165	·77	32·2
4	390,000,000	2,103,000	190	·80	34·0
5	441,000,000	2,187,000	201	·82	35·2
6	487,000,000	2,255,000	216	·835	36·1

The figures in the fifth column are those given in

CHAP. II., *Art. 1.* The figures in the last column show the corresponding available falls, after deducting the loss of 14 inches. It will be seen that, owing to this deduction, the available falls for the shorter periods are reduced in a greater ratio than the figures in the fifth column.

In arranging for the supply of towns in the British Isles it is usual to design the reservoirs so as to equalise the flow of the three driest consecutive years. Existing reservoirs, old and new, usually contain from 140 to 170 days' supply, but some contain less. The above table shows that for the assumed fall of 60 inches and loss of 14 inches, the capacity of a reservoir, to allow for a six-year dry period, has to be 49 per cent. more than for a three-year dry period, while the daily supply from it is only 13 per cent. greater.

The following statement gives Deacon's figures for mean annual rainfalls ranging from 30 to 100 inches. The columns marked R show the reservoir capacities in millions of gallons, and those marked S the daily yields of the reservoirs in thousands of gallons. The figures for other falls can be interpolated. For a fall of, for instance, 50 inches, the figures, whether of R or S, are practically a mean between those for falls of 40 and 60 inches.

Number of Years whose Supply is to be Equalised.	F=30.		F=40.		F=60.		F=100.	
	R.	S.	R.	S.	R.	S.	R.	S.
1	35	300	79	695	166	1475	345	3040
2	85	470	140	900	258	1815	495	3600
3	120	560	190	1050	329	1987	610	3900
4	150	620	230	1110	390	2103	710	4100
5	175	650	260	1170	441	2187	800	4230
6	195	680	290	1220	487	2255	887	4320

In all cases the loss is supposed to be 14 inches annually. If it is 15 or 13 inches, the reservoir capacity is less or more by about five, ten, or fifteen million gallons, according as the number of years in column 1 is 1, 3, or 6. And the daily yield is less or more by about 50,000 gallons.

With a low rainfall the advantage of a large reservoir is somewhat increased. The capacity of the six-year reservoir for a fall of 30 inches is 63 per cent. more than that of the three-year reservoir, but the supply is 22 per cent. greater.

The figures given above for reservoir capacities are suitable for the British Isles. They assume that the distribution of the rainfall is the least favourable that is at all likely to occur. Deacon states that the figures do not relieve the engineer of the exercise of judgment. As regards the British Isles, the chief questions on which judgment has to be exercised are whether to equalise the flow of three years or of another number, and how much to allow for loss. As already stated, three years is the period usually taken. The figures are suitable for most places in Europe, but in some places, *e.g.* on the Mediterranean coast, the distribution of the rainfall is somewhat less favourable than in the British Isles. In other parts of the world, and notably in or near the tropics, the distribution of the rainfall must be specially studied, and a diagram be prepared on the same principle as in the case of fig. 58. The diagram should be extended to cover the desired number of years. In hot countries loss by evaporation from the surface of the reservoir should be allowed for. In India during the hot dry months this loss may be half an inch in twenty-four hours.

In the article above quoted it is shown that if, as commonly happens, the consumption of water is, in summer, greater than the mean, and in winter less, the conditions are still more trying for the reservoir; and that in the case where the summer consumption is 13 per cent. greater than the mean, the capacity of the reservoir which impounds the water of the driest year must be 33 per cent., instead of 30 per cent., of the total supply impounded during the year. It would then contain 121 days' instead of 110 days' mean supply. In the table on page 170 the number of days' supply is 113. From this it appears that the tables from which extracts have been given are calculated on the basis of a constant consumption. This, however, in the case where the number of years whose supply is equalised is greater than one, makes, owing to the increased size of the reservoir, no practical difference.

The calculations for the great reservoirs in Radnorshire for the supply of the city of Birmingham are as follows (*Min. Proc. Inst. C.E.*, vol. clxxx.). The ratio of the mean fall in the three driest years to the mean annual fall was taken as ·80 instead of ·77. There is some difference of opinion as to the best figure:—

Mean annual fall determined from readings of various gauges . . .	65 inches
Mean fall of three driest years . .	52 ,,
Deduct loss from evaporation and absorption and losses during floods . .	15 ,,
Available rainfall . . .	37 ,,

This multiplied by 44,000 acres, the area of the catchment, gives 102 million gallons per day. Of this, 27,000,000 gallons is compensation water, leaving

75,000,000 gallons for Birmingham. Capacity of reservoirs, 17,250,000,000 gallons, or 169 days' supply.

3. **Earthen Dams.**—Before an earthen dam is made, any soft soil on the site should be removed and the ground downstream of the site should be drained. A

FIG. 59.

few trenches, running parallel to the axis of the dam, can be dug so as to give the dam a hold, though there is never any danger of its being moved horizontally by the thrust of the water. If the ground has a sidelong slope it should be benched as shown in fig. 59. The front slope of an earthen dam is generally about

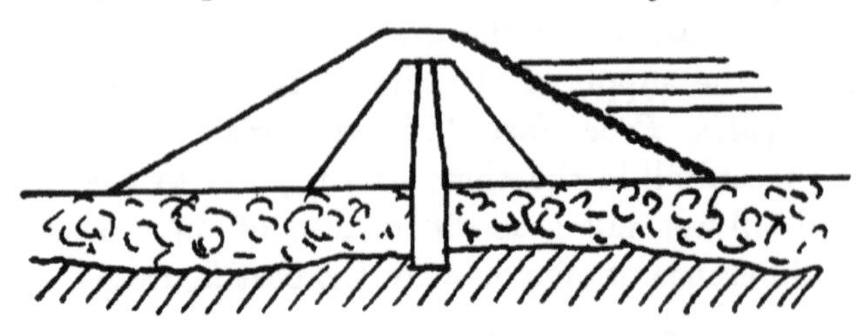

FIG. 60.

3 to 1, and the rear slope about 2 to 1. The top has a width of $\frac{1}{3}$ to $\frac{1}{2}$ the greatest depth of water held up, and is 5 to 10 feet above the highest water-level. The borrow pits from which the earth for the dam is got should not be near enough to it to in any way affect its stability.

In England, and generally in other countries, an earthen dam has a core-wall (fig. 60) which is carried down to an impervious stratum, and is keyed into it to a depth of a foot or more in the case of hard rock and several feet in the case of clay. On this core-wall the impermeability of the dam chiefly depends. The core-wall may be of clay puddle, concrete, or masonry. In England it is generally of clay puddle. The core-wall sometimes extends down to a depth of 100 or 200 feet. Its top is horizontal and about level with the highest water-level. It is desirable not to make the foundation stepped, but to let it follow the profile of the impervious stratum. The wall is keyed at its ends into the sides of the valley or gorge. A core-wall of concrete or masonry is, in a high dam, necessarily a comparatively thin structure, and it may be subjected to great strains by unequal pressures of the earth which surrounds it. It is therefore to some extent liable to crack. A core-wall of concrete used for the water-works of Boston, U.S.A., is 100 feet high, 8 feet thick at the base, and 4 feet thick at the top. A clay-puddle wall, being plastic and moist, at least during the period immediately succeeding the construction of the work, is not very liable to crack. The top width of a puddle wall may be 4 to 10 feet, and the batter of the sides from 1 in 20 to 1 in 8. The clay used for the wall above the ground-level should contain about 33 per cent. of sand and stones. This diminishes its shrinkage if it dries. It should not be given too much water in mixing. It should be thoroughly mixed and worked up and trodden down.

The clay puddle and the earth of the dam should be carried up uniformly. The allowance for settlement may be $\frac{1}{30}$ to $\frac{1}{50}$. The earth should be deposited in thin layers, moistened and rammed, and all clods broken.

In India and some other countries, instead of the earth being rammed, cattle or sheep are driven over it repeatedly. This makes earthwork of most excellent quality, and the settlement, if any, is very small.

In cases where, owing to a fissure in the rock below the bottom of the puddle trench, water comes through under the puddle, it is usual to carry it away in a pipe running vertically in a groove up the side of the trench and then horizontally till it emerges from the dam. Such water, and any other leakage, can often in Great Britain be used as part of the compensation water. There is, however, a certain chance, when there are water-bearing fissures in the rock below the bottom of the trench, that some percolating stream of water may wash away the puddle, and it is preferable to use a concrete core-wall in such cases, carrying it up to about ground-level and keying it into a much thicker wall of puddle which is carried up to the water-level.

It has been suggested that the clay puddle or other impervious layer should be placed, not vertically and in the middle of the dam, but lying on the upstream face of the dam, so as to keep out water from the whole dam instead of from only half of it. Objections to this, if clay puddle is used, are that vermin may bore holes in it, and that, with some clays, it would slip. These objections might be overcome to some extent by laying a pitching of concrete blocks over the puddle. Other objections, applying also when masonry or concrete is used, are that the superficial area and cost are increased, and that cracks would occur from settlement of the earth and from changes of temperature when the water in the reservoir was low. A good many cases have occurred in which an impervious layer laid on the slopes has failed from one cause or another. In France it is usual to rely on such a layer—concrete is used—and to dispense with a core-wall. The practice of having a vertical wall appears to be the best, and is the most widely adopted. When puddle is used the weight of the mass above it forces it to completely fill the trench, and when once it is in position and covered up it is not at all likely to be damaged.

The outer portion of a high embankment sometimes slips (fig. 61), and precautions should be taken against this. A slip may occur if the site of the dam has not been carefully selected as to geological formation, or if

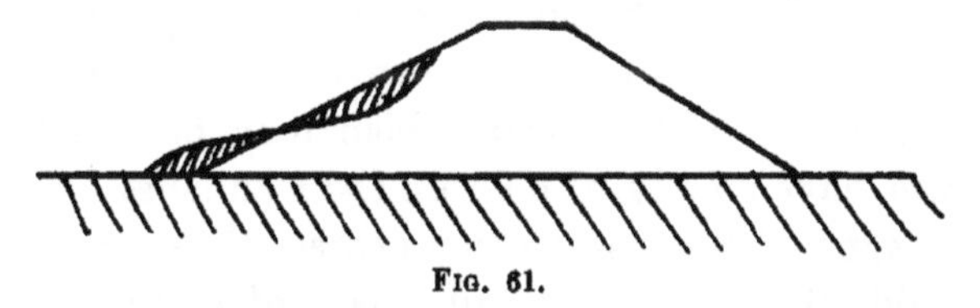

FIG. 61.

there is unequal settlement owing to the work having been done at different times. One cause of slips is sudden and partial changes in the degree of saturation, and another cause is excessive saturation. Some clays when wet require extremely flat side slopes, and will not stand even at 5 to 1. The outer parts of the embankment are not required for stopping percolation (this will be further considered in the following paragraph), and, though they must be carefully laid and consolidated, they should be of porous material, and the part on the downstream side of the dam must be well drained. A series of surface drains may be arranged and filled with loose stone and gravel. There is also a distinct advantage in using heavy material such as small stone for the lowest portion of the outer parts on both sides of the dam.

When good material cannot be obtained, the side slope on the downstream side of the dam may be flattened. A side slope starting at the top with 3 to 1 and becoming 4 to 1 lower down, and finally 5 to 1 at the base, is a very good form for prevention of slipping and generally for the safety of a dam. The part on the side next the reservoir is not likely to slip. It becomes soaked, but it has the pressure of the water against it and is pitched. In Madras, where reservoirs are very numerous, the slope on the side next the water is generally only 1½ to 1.

The different parts of an earthen dam fulfil two distinct functions. Some parts, which may be called the staunches, have to stop the percolation of water from the reservoirs. Other parts, which may be called the supports, have merely to hold up the staunches. In the British type of dam the portion nearest the core-wall on either side (fig. 60) is generally made of earth specially selected for impermeability. The distance to which it extends from the wall depends partly on the quantity of such earth available. In any case it has to be very carefully made and consolidated, to avoid unequal pressures on the core-wall, or unequal settlement which might cause it to part from the wall. One of its functions is to keep the core-wall moist when the water-level in the reservoir falls. Whether it is also to be considered as a staunch or a support might at first appear to be of no consequence, but it is of importance as affecting the question of drainage. The support on the downstream side of the dam must, as has just been seen, be made of porous material and be well drained; but obviously a staunch must not be porous, nor can it be penetrated by drains. The question must be decided in each case according to judgment. In a discussion which took place on the above-mentioned paper by Strange, at the Institution of Civil Engineers, much diversity of opinion was expressed among eminent engineers as to the desirability of draining the downstream half of the dam, *i.e.* the part downstream of the core-wall. By some it was urged that drainage is necessary to lessen the chance of the earthwork slipping. Others contended that any drain which penetrates the dam must facilitate the percolation of water from the reservoir. It is clear that some of the speakers regarded the dam downstream of the core-wall as being partly staunch, and some as being wholly support. If for any reason there seems a chance of water leaking through the core-wall, it is desirable to regard the earth-filling next to it as staunch.

In Western India a kind of puddle is made by mixing three parts of "black cotton soil" with two of sand. The object of the puddle wall is only to prevent water from finding its way along the surface of the ground. It is carried down only to a fairly water-tight stratum and is carried up only to 1 foot above the ground. Above that the mass of the dam is made of black cotton soil as a staunch, with more porous material on both sides of it.

In order to afford full protection against waves and their splashes, the pitching on the upstream face of a dam should extend up to a height of 5 feet, measured vertically, above the highest water-level. In the case of a dam in which the "fetch" or distance over which the waves have been in process of formation exceeds two miles, the above height should be slightly increased.[1]

[1] The height of a wave is supposed to be $1{\cdot}4\sqrt{\text{fetch}}$, but this allows nothing for splashing.

The pitching is usually of stones roughly squared at their outer ends and laid on a layer of broken stones.

The water from a reservoir is usually drawn off by means of pipes which are laid inside a masonry culvert built under the dam. The pipes can thus be inspected. The culvert is blocked at its upstream end by a thick masonry wall through which the pipes pass. Accidents which have happened in the past have been due to weakness of the culvert or to water finding its way along the outside of the masonry. The culvert can be made of proper strength, and it should have a thick coating of clay puddle which is worked into the clay-puddle core-wall of the dam. If the core-wall is of masonry or concrete, the masonry of the culvert is properly joined to it. In many cases the culvert and pipes are taken through a cutting or tunnel and not under the dam.

At the upstream end of the culvert there is a masonry tower—access to it is obtained from the top of the dam by a foot-bridge,—and from it valves for opening and closing the pipes are worked. If the reservoir is for the water supply of a town, it is arranged, by means of a vertical pipe, that the draw-off can be at various levels so that the surface-water can always be used. In the case of some of the towers at the reservoirs whence Birmingham is now supplied, the vertical pipe consists of a number of steel cylinders with gun-metal faces which are so accurately made that the joint is water-tight when one cylinder merely stands on another. The draw-off is obtained from a given level by lifting a particular number of cylinders. Sometimes the tower is made of reinforced concrete. When it is lofty it should be strong enough to resist a strong wind, blowing when the reservoir is empty.

4. **Masonry Dams.**—For heights much exceeding 110 or 120 feet a masonry dam may be cheaper than an earthen dam ; and in case a flood occurs while work is in progress the masonry might suffer little injury, while earthwork might be swept away completely. Masonry dams are usually built of random rubble masonry with faces of dressed stone. Such masonry weighs about 140 lbs. per cubic foot, and is ordinarily quite safe when subjected to pressures of 20 tons per square foot, but in a masonry dam a high factor of safety is necessary, and 15 tons per square foot may be allowed. In a wall of such masonry with both faces vertical, the pressure, owing to the weight of the wall, will reach the above limit when the wall has attained a height of about 220 feet.

In a masonry dam, although the masonry is always of the best quality, it is a rule to calculate the dimensions so as to give no tension on any part of the masonry. Any crack or opening of a joint, occurring perhaps before the masonry had hardened, would let in water, and its pressure would tend to gradually extend the crack and eventually to overturn the portion above the crack.

Fig. 62 shows the upper part of a masonry dam. The lines with arrows show the vertical force due to the weight of the masonry above A B, the horizontal force due to the water-pressure on it—acting at two-thirds of the depth,—and the resultant of these two. In order that there may be no tension on the masonry, the resultant must always fall within the middle third of the thickness of the dam. In order to prevent its falling outside the middle third, the downstream face must be splayed out, and the splay will go on increasing somewhat. Suppose, now, that the reservoir is laid dry. It will be found that in the case of a dam more than 100 feet high the pressure due to the weight of the wall alone will fall outside the middle third—to the upstream side of it, of course—of the thickness of the wall, and a slight splay must be given to the upstream

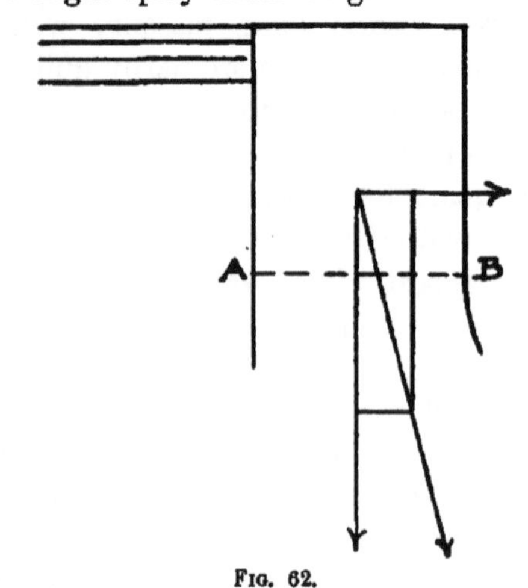

FIG. 62.

side. The vertical pressure of the water on this splayed part must be taken into consideration. The limit of pressure, 15 tons per square foot, may eventually be reached owing to the height of the dam, and additional splay may have to be given for this. When the outside splay becomes considerable a further allowance is made for it, because the stress at the edge of a horizontal section is tangential to the face. In order that the tangential stress may not exceed 15 tons per square foot, the vertical stress at the outer edge of a horizontal section of the dam must not exceed about

12 tons. By following the above rules the section of the dam can be calculated, beginning from the top and working downwards. The resulting profile of the dam is somewhat as shown in fig. 63. If a masonry dam is designed on the principles given above—that is, so as to be safe as regards crushing and overturning—it will be safe as regards shearing or sliding horizontally, but a test calculation can easily be made for this.

Calculations of the above kind do not, of course,

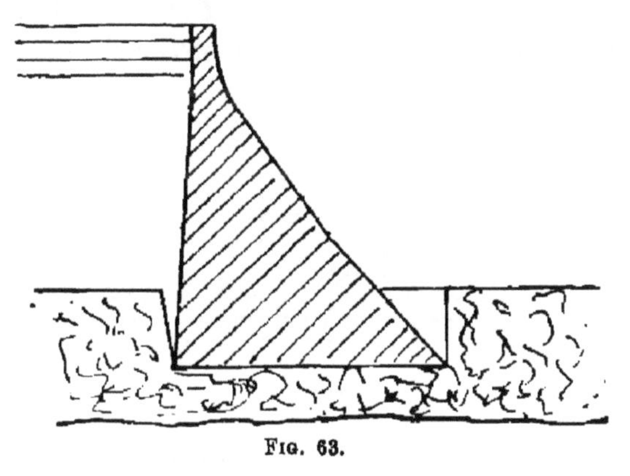

FIG. 63.

enable all the stresses in a solid mass of masonry to be found. Great stresses are caused by expansion and contraction owing to changes in temperature. Others are caused by the connections of the dam with the rock on which it rests and with the sides of the gorge. The method of calculation described above indicates a suitable form for the profile of a dam. The large factor of safety adopted allows for other stresses. The sections of the oldest dams, made in Spain, were somewhat as shown in fig. 64, and contained about twice as much material as was necessary. The object of the calculations is to save this needless expenditure.

Masonry dams designed on the above principles have been constructed for heights ranging up to nearly 300 feet, measured from the foundation to the top. The foundation is always on hard rock free from fissures. Generally a foundation trench is cut. The ends of the dam are carried into the rock on the sides of the gorge. They should not, however, if the sides of the gorge are steep, be built in with mortar, but be allowed to expand and contract vertically, a water-tight joint being made by means of asphalt (*Ency. Brit.*, Tenth Edition, vol. 33, "Water Supply"). This obviously reduces the

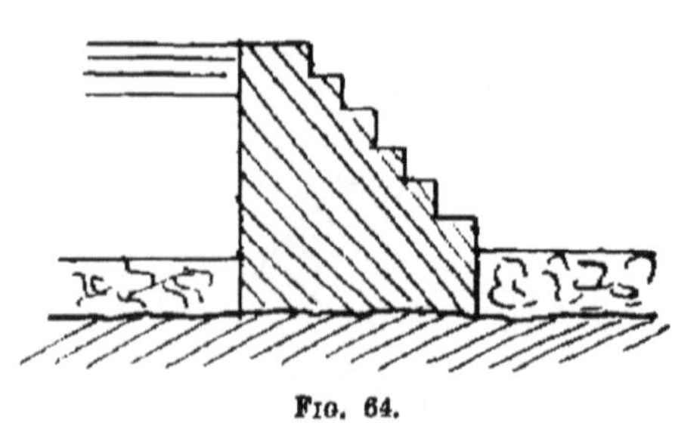

FIG. 64.

straining. A dam should be built in cool weather, so that any stresses to which it will eventually be subjected owing to changes in temperature will be chiefly compressive. The upstream face should be as water-tight as possible. There should not, however, be too sudden a change in the character of the masonry from the face work to the inside work. If there are any springs, they must be carefully connected to pipes and carried outside the dam. No water must be permitted to get under or inside the dam, either from springs in the sides of the gorge or from the water in the reservoir. Many existing dams leak slightly where they join the sides of the valley, and most have developed some vertical cracks normal to the face.

Out of some hundred high masonry dams which have been erected, only three are known to have failed. Of these, the Puentes dam was partly founded on piles; and in two, the Habra and Bouzey dams, the rule of the middle third was not attended to. Another dam, not so high, the Austin dam, in Texas, U.S.A., failed seven years after construction. It was 65 feet high and founded on limestone, the width of the base being 66 feet. Springs in the bed and sides of the gorge had, during the construction of the dam, given much trouble, and had, after its completion, forced their way through the underlying rock. At the time of failure 11 feet of water was passing over the dam, which sheared in two places, a length of 440 feet of it being pushed forward for 40 or 50 feet without overturning, but subsequently breaking up. The dam was founded in a trench cut in the rock. The rock on the downstream side of the foundation trench appears to have been worn away by the water, so that there was no longer a trench (*Scientific American*, 28th April 1900). The above, however, does not seem to be sufficient to account for failure. The horizontal water-pressure on a 1-foot length of the dam would be 180,000 lbs. and the weight of masonry to be moved perhaps 320,000 lbs. It seems probable that water from upstream found its way under the dam and exercised a lifting force on it and so caused it to slide.

If a masonry dam, instead of being straight, is made curved on plan, with its convexity upstream, it acts as an arch, and its thickness can, in the case of a fairly narrow gorge, be greatly reduced. This type of dam is a suitable one to use when the sides of the gorge are of firm and solid rock and there is no doubt about their being able to stand the thrust without yielding. Several dams of very considerable size have recently been built in this way. The thickness of the upper part of the dam and the ratio of the versed-sine of the arch to the span can be decided on by the methods used for arches in general. The lower part of the dam is made thicker. The lowest part cannot act as an arch, because it is attached to the foundation. It is, however, assisted by the portion above it, which acts as an arch, and thus need not be so thick as in a "gravity" section. The Bear Valley dam, which is 64 feet high, is only 3 feet thick at the top. The thickness increases gradually to 8½ feet at 48 feet from the top. The chord of the curve is 250 feet and the radius of curvature 335 feet. If the gorge is wide, the thickness of the arch comes out so great that nothing is saved by adopting the curved form. But in such a case, and in any case, a dam can be made slightly curved so as to offer a greatly increased resistance to overturning. It need not act as an arch, and can be prevented from so acting, in order that excessive stresses may be avoided, by letting the ends of the dam, after they have entered the grooves cut in the sides of the gorge, stop short of the ends of the grooves.

During the last few years much attention has been given to the investigation of the stresses to which a masonry dam is subjected. Some investigations have been theoretical and others practical, models of india-rubber and other substances having been used for experiment. The investigations show that generally the stresses in a model of a dam are very much the same as would be expected, but that there is a tensile stress, previously overlooked, near the point M (fig. 65), where the dam rests on its foundation. The tension is on the foundation, on the line M N, and is due to the horizontal thrust of the water. It is natural that in an elastic model this stress should manifest itself by deformation. In the case of an actual dam resting on rock, matters are different; but this tensile stress deserves consideration. For the present let it be supposed that

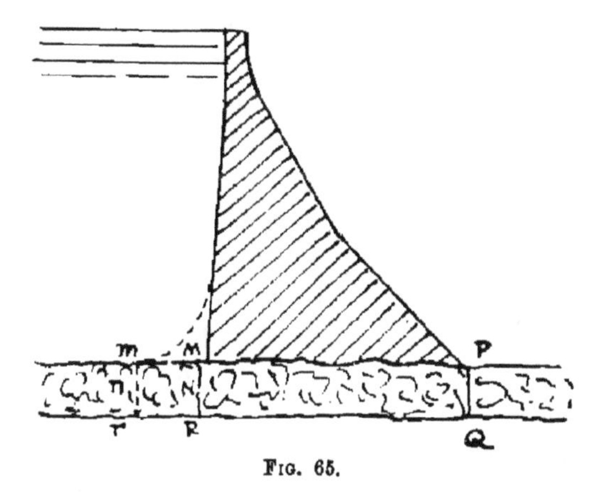

Fig. 65.

there is no trench, the dam merely standing on the rock. Suppose that the rock has only the thickness M R. There is tension in M N, and probably compression in N R. It is assumed that, along the base M P, there is perfect union between the dam and the rock. The tension to which the rock is occasionally subjected owing to changes of temperature may exceed any tension due to the water-pressure, but it is conceivable that the tension occurring from both causes might cause a crack at M N, and that this might extend to R. This implies a minute sliding of the dam and of the rock below it, movement taking place on the plane R Q. The thrust of the water is now resisted by the rock downstream of P Q. The dam, with the rock M R Q P adhering to it, tends to rotate about the point P. The tendency to rotate will be enhanced if water enters M R, and still more if it enters R Q. No rotation can, however, take place unless the rock at M R is splintered away. The rock would also have to fracture at P Q. It has been suggested that the upstream face of the dam be made curved as shown by the dotted line. This would shift the chief tension to *m n*, and the dam, with the rock beneath it and the weight of the water above the curved portion, would obviously offer an increased resistance to rotation about P. The cost of the dam would of course be increased. The danger of a crack forming at M N seems to exist only when there is a thin upper stratum of rock not firmly connected to rock below. When this condition is believed to exist, a masonry dam, if built at all, should have the upstream face curved as above described. In the case of any existing dam of great height, when the above condition is suspected to exist, the reservoir might be laid dry, and if any crack at M N is discovered a curved portion could be added; but in this case the union between the new and the old work would be imperfect, and the curve should start from high up on the upstream face of the dam. It has been suggested that asphalt or some impervious material be laid on the rock to prevent water from entering any crack. It would, however, not only have to be laid upstream of the dam, but to extend under part of the dam, and thus weaken it to some extent.

In the case (fig. 63) in which the dam is founded in a deep trench, the building up of the upstream triangular space and uniting the material both to the dam and to the side of the trench, might be of some use, but a crack might form in it. It would be desirable to add a curved portion, as above described, on the top of the rock if sound, or to remove the unsound rock and widen the trench and then add the curved portion. Adding material to the downstream triangular space, and uniting it well, would also increase the resistance of the dam to overturning, not so much because of the additional weight, as because of the raising of the point about which the dam would have to revolve in overturning.

Several recent dams have been built of cyclopean concrete, blocks of rock as heavy as 10 tons being sometimes used in the work. Such blocks are laid on one of their flat faces. In the U.S.A. some reservoirs have been made with walls of reinforced concrete, backed by earth embankments (*Min. Proc. Inst. C.E.*, vol. clxxxix.), and also of cyclopean masonry reinforced with steel rods. Another kind of dam which has been used in the U.S.A. is the rock-fill dam with a core—corresponding to the puddle wall in an earthen dam—of steel plates riveted together and made water-tight and inserted into the rock at each side. In the case of the East Canyon Creek Reservoir, Morgan, Utah, the dam is 110 feet high. The steel plates vary in thickness from ⅜-inch at the bottom to ¼-inch at the top, and are embedded in asphaltum concrete and rest on a concrete base. The dry-stone work of the dam is hand-packed on both faces, and also on both sides of the core. The rest is thrown in. The upstream face is 1 to 1, and the downstream face 2 to 1. The waste weir is at one end of the dam and is continued by a flume, so that the water falls clear of the dam. The outlet is a tunnel in the rock.

CHAPTER XIV

TIDAL WATERS AND WORKS

1. **Tides.**—The tides or "tidal waves" are caused by the attraction of the moon and the sun. The phenomena are complex, and a full discussion of their causes need not be given here. When the tide rises it is said to "flow," and it is called the flood tide; when it falls it is called the ebb tide. The period between one tide and the next, *e.g.* from high water to high water, is about twelve hours, twenty-five minutes. At a spring tide the range of the tide is greater than usual; at a neap tide less. Where there are channels, as, for instance, the seas which surround the British Isles, the tidal waves run up them as the tide rises in the neighbouring ocean, and run back as it falls. At some places, as Southampton, the tide comes in from two directions, and there is a double tide. The times and levels of high and low water at various places have been ascertained by observation, and are recorded. The levels are, however, liable to be affected by winds. A wind blowing towards the shore raises the level of both high and low water; a wind blowing off shore lowers both levels. A severe storm in the North Sea has caused a double tide at London Docks, by accelerating the North Sea tidal wave.

In a funnel-shaped estuary, especially if it faces the direction of the tidal wave in the sea, the tide in going up the channel increases in velocity, and the momentum of the water causes it to rise higher and higher as the width decreases. At the upper end of the Bristol Channel the range of the tide is double the range in the sea outside the channel. The Bay of Fundy is another place where a similar phenomenon occurs. When a river or estuary is shallow and the range of the tide is great, so that its rise is rapid, the flood tide in some cases advances in the form of a wave or "bore," causing a sudden rise in the water-level and a sudden reversal of the flow of the stream. A bore is most pronounced at spring tides. That of the Severn is well known.

In the case of a tide running along a coast or up an estuary, the water of the flood tide, after it has ceased to rise, continues for a short time, owing to its momentum, to flow in the same direction as before. The same thing happens when the ebb tide ceases to fall. The tide also acquires special velocity, just as a river does, round any projecting headland.

The rise and fall of the tide are least rapid near the turns of the tide. If the time from the beginning to the end of the flow be divided into six equal parts, the proportional rise of the water will be approximately as follows. And similarly with the fall during the ebb.

Time	1	2	3	4	5	6
Rise of water	·067	·25	·5	·75	·933	1

Tidal waters are frequently charged, more or less, with silt, obtained from the shore or from shallows near it, either by currents or tidal waves sweeping along it, or by the action of ordinary waves. Tidal waters flowing up and down the lower portions of rivers render them to an enormous degree more capable of carrying navigation and, especially if they become enlarged and form estuaries, more capable of being altered by training works.

A tide-gauge is constructed on the same principle as a self-registering stream-gauge. The rise and fall of the water are reduced, by mechanism, to a convenient range, and are recorded on a band carried on a drum, which is caused to revolve by clockwork. Another kind, which depends on the use of an inverted syphon filled

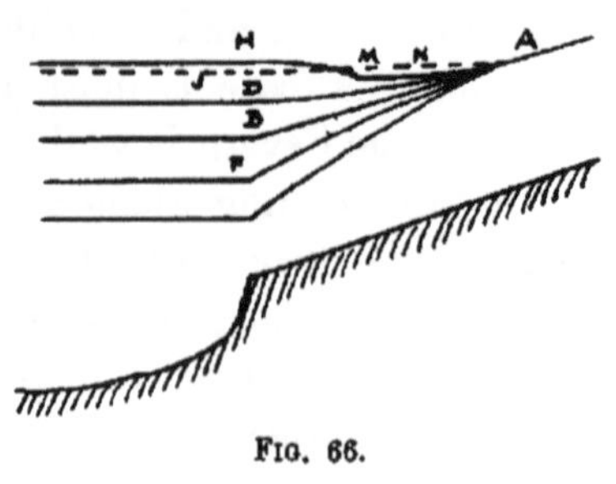

Fig. 66.

with air and a syphon of mercury, is described in *Min. Proc. Inst. C.E.*, vol. clxiv.

2. **Tidal Rivers.**—Let A B (fig. 66) be the surface of the lower part or mouth of a river, supposed to be of uniform width, and let B be the mean sea-level. As the tide rises to D the water of the river is headed up and assumes the line A D. When the tide falls to F there is a draw, the river surface taking the line A F. If the rise of the tide B H is so great that the discharge of the river cannot keep pace with it, so as to fill up the whole space between A and H to the level of H, there will be a flow of sea water from H to some point **M**, and of river water from A to M. The point M will be

lower than A and H. If the tide now turns and the water-level H begins to fall, there will still be a flow along HM. For a brief period it will be due to momentum, but it will continue until, by the rise of the water-level at M and the fall at H, the surface has assumed the form indicated by the dotted line ANJ. While this is happening, the point corresponding to M —where the concave curve of the upland water meets the convex curve of the tidal water—rises higher and shifts seaward. The character of the two curves remains the same, but they become flatter and the surface NJ nearly level.

Thus the time of high tide at M is later than at H. It is later for each point passed in going up the river from H towards A. Eventually a point A is reached where there is no tide, that is, no rise or fall. Far below this point, between A and B, there is a point above which there is no upward current but only a slackening of the downstream flow. At H a diagram showing the rise and fall of the tide is symmetrical, at N the rises and falls are less than at H, and the periods of their occurrence later. In going up the river the duration of the flood tide decreases and that of the ebb tide increases. The flood tide attains its greatest velocity soon after its commencement, the ebb tide towards its close. The distances to which the tidal influence extend are of course greater the greater the range of the tide and the flatter the slope of the river. The discharge of the river of course varies. The greater the discharge the more the rise of the river tends to keep pace with that of the tide and the less the distance to which the tidal influence extends. On a longitudinal section of the river, the high-water line will be shown as

A N H. This is merely done for convenience. It is never high water at all points simultaneously. To show the actual state of affairs at various stages of the tide, series of lines must be drawn as in fig. 67, where the firm lines show the flood, and the dotted lines the ebb tide.

The flow in the tidal reach of a river is the same as if the water was alternately headed up by a movable weir and then allowed to flow freely and be drawn down. If the water carries silt, the tendency for deposit to occur is (CHAP. V., *Art. 2*) no greater than if there was no heading up or drawing down. The tendency depends

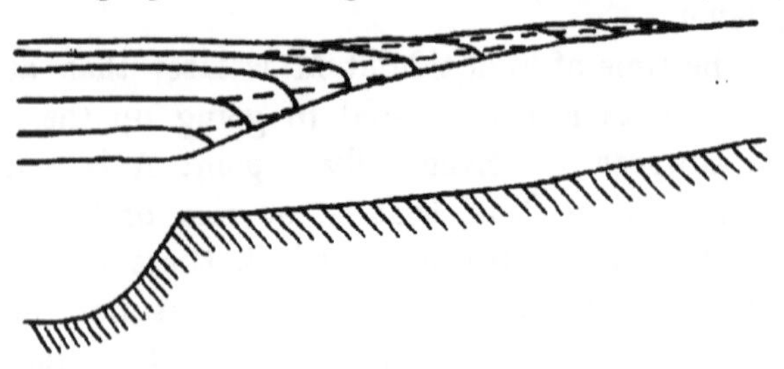

FIG. 67.

chiefly on whether there is, on the average, any reduction in velocity or increase in depth as compared with the non-tidal upstream reach, and whether the water in that reach is fully charged with silt. If both the answers are in the negative, no deposit due to river silt is likely to occur in the tidal reach.

If the sea water is charged with silt, it will of course carry silt into the river as it flows up, but the whole volume of water which enters has to flow out again. On the whole, the tendency for silt to be deposited in the river is due only to the period of "slack tide" near the time when the flow ceases. The tendency is seldom marked.

If the sea water carries silt and the river water is clear, the latter assists of course in removing any deposit —that is, it tends to keep the channel clear.

If the river channel is soft and if the sea water carries no silt, it may, in passing up and down the river, become charged with silt and return to sea still carrying it. It thus has a scouring effect on the channel, and may deepen or widen it. If, owing, for instance, to the flattening of the bed slope in its lower reaches, the river tends to deposit its own silt in its tidal reach, the sea water may prevent this deposit. Thus, as regards silting in the tidal reach of the river, the tidal water of the sea has little prejudicial effect if it is silt-laden, and a beneficial effect if it is not. Silt is likely to deposit in the tidal reach of a river of uniform width, only in a case in which the river water carries much silt, and the slope is flat or cross-section great compared to that of the upper reach.

Sea water is heavier than fresh water by about 2·4 per cent., and this, to some extent, prevents their mixing. At all stages of the flood tide the tendency at the point where the fresh water meets the salt water is for the fresh water to accumulate towards the surface and the sea water towards the bottom. When the tide begins to flow up the river there may be a low-level salt water current moving landward and a high-level fresh water current moving seaward, but this is quite a temporary state of affairs. The surface slope is landward, and the water moving seaward is not moving in obedience to the surface slope. It is only moving as a result of momentum previously acquired. The low-level current may have some extra velocity and extra scouring power, but this cannot be much, because the mean landward velocity of the whole stream must, owing to the internal resistances caused by the two currents, be less than it would be if there were not two currents. Moreover, the state of affairs is temporary. The two kinds of water mix eventually, and their temporary separation has no considerable effect on the general tendency of the river in the tidal reach to scour or to silt.

A body of water included at any moment between any two cross-sections of the tidal portion of a river may not reach the sea during the next ebb tide. In this case it will flow back up the channel with the next flood tide, and so be kept moving up and down, getting nearer, however, to the sea at each tide.

De Franchimont has shown (*Min. Proc. Inst. C.E.*, vol. clx.) how a diagrammatic route-guide can be prepared for any tidal river to show pilots or captains of vessels the best times for starting on voyages up or down the river, and for passing each point on it.

3. **Works in Tidal Rivers.**—If any works are required in the tidal portion of a river, the principles to be followed in designing them are the same as if the river was non-tidal. All that has been said in CHAP. VIII., *Arts. 1* to *3*, applies to them. The river may be straightened or trained or dredged. Generally training and dredging are combined. Any dredging in the portion of the river nearest the sea will not, of course, alter the water levels near the mouth, but it will alter them further up. The tide will come up in greater volume and will rise higher and extend further up. The ebb will be facilitated, and the low-water level will be lowered. If any narrowing of the channel near its mouth is effected by training walls for the purpose of lowering the bed, the effect on the volume of tidal water

entering the river must be taken into consideration. If the narrowing is confined to a reach near the mouth, and if the resulting deepening is not sufficient to counteract the effect of the narrowing, the volume of tidal water reaching the unnarrowed portions of the channel will be reduced, and this may be injurious. Its scouring action may be insufficient. The proper course may be to continue the narrowing upstream. If this is done, then it is obvious that the width of channel in which deep water is to be maintained at high water, or which is to be kept free from deposit, is reduced in about the same proportion as the volume of tidal water is reduced, and no harm is likely to result.

Any weir or similar structure which abruptly stops the flow of the tide up a river checks it of course for a long distance back, perhaps to the mouth. Old London Bridge used to obstruct the tide, and its removal increased the range of the tide, and was beneficial.

Tidal rivers generally widen out to some extent near their mouths, and are thus rather estuaries than rivers. The works in such rivers are more fully discussed in *Art. 6.*

4. **Tidal Estuaries.**—If, instead of a river of uniform width, there is an estuary whose width increases steadily towards the sea so that it is funnel-shaped, the conditions described in *Art. 2* are modified. An estuary is formed first by the waves of the sea, which wear away the angles at the mouth of the river and allow the tide to enter in greater volume, and then by the flow and ebb of the tides. The slope of the bed of the estuary is usually much flatter than that of the river, and the water surface is as shown in fig. 67. The tidal movements extend further upstream than in the case of a river, not

only because of the greater difficulty experienced by the upland water in filling up the wide channel of the estuary, but because of the momentum of the tidal water driving its way up the funnel-shaped channel (*Art. 1*). The capacity of the estuary is of course much greater than is required for the discharge of the upland water alone. If the sea-level remained always at one height and if the upland water contained silt, it would tend to deposit in the estuary and would certainly deposit in it to some extent. The action of the sea water is the same as described in *Art. 2*, scouring if it is clear when entering, of less account if it is not clear. Owing to the funnel shape of the estuary, the tide rises higher at its upper end than if the estuary were replaced by a river channel, and the tide also extends further up. This may partly or wholly compensate for the greater tendency of silt to deposit in an estuary as compared with a river channel.

The ebb tide in an estuary does not always follow exactly the same course as the flood tide. Of course the lowest parts of the estuary are filled first and emptied last, but the channels are not all continuous. A channel open at its lower end may have a dead end at its upper termination, and *vice versa*. Also, at sharp bends in the channels, the momentum of the water may cause differences in the paths traversed by the flowing and ebbing currents. Wherever there is a deep channel the water from the adjacent sandbanks tends, towards the close of the ebb, to flow cross-wise into the channel, and in doing this it to some extent washes down the banks into the channel.

5. **Works in Tidal Estuaries.**—Estuaries, when shallow, offer great facilities for training. It used at

one time to be said that any change which reduces the volume of tidal flow must be injurious. It would be injurious to restrict the mouth of the estuary, unless it were exceptionally wide, and leave the rest untouched. If the whole estuary is narrowed, and a suitable funnel shape preserved, the width to be kept open is, relatively to the size of the mouth, no greater than before, and the tide may flow up as far as before, and rise to as high a level. The narrowing, if properly arranged, will improve the shape of the estuary and cause an increased scour. The effect of the upland water is also greater in the narrower channel. Improvements to estuaries are not, however, restricted to training. There is always one or more deep channels, and the best of these can be selected and improved by dredging. The channel should be one along which both the flood tide and the ebb tide will run. The above remarks as to training do not apply to a case in which there is a bar outside the mouth of the estuary. Training might check the scour at the bar. Bars are treated of in CHAP. XV.

If an estuary is not funnel-shaped, if, for instance, it widens out very rapidly, the tidal flow is much less effective in keeping the channel open. In this case, training works, which would give the necessary funnel shape, are indicated rather than dredging. If an estuary is narrow at the entrance, the flow is much less powerful, unless the narrow part is of greater depth, but even then the force of the tide is reduced owing to the change in the shape of the channel.

The bed of an estuary may be of such soft or sandy material that a dredged channel would be likely to be quickly filled up again by the slipping in of material at the sides (*Art. 4*). In such a case an untrained channel can only be kept open to its full depth by constant dredging, and probably the best course is to construct a trained channel, although it may be more expensive than in the case of a harder channel, because of the depth to which the foundations of the walls must be sunk into the soft bed. Also, if the bed of the estuary is constantly shifting, a dredged channel alone will not succeed, and training must be resorted to. Again, the bed may be of such hard material that training walls would not cause it to scour. In this case a channel should be dredged and need not be trained.

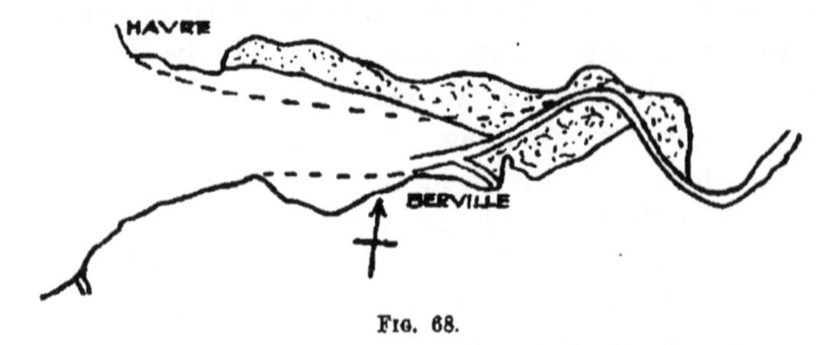

FIG. 68.

For the great body of intermediate cases in which the deep channel can be formed either by dredging or training, both methods can be adopted. A common plan is to train the upper part and to dredge the lower part where the estuary is wider and the training walls would be more exposed to the waves.

When an estuary is thus partly trained, the deepening due to the training does not extend far beyond the point where the walls terminate. The deposit of material along the sides of the estuary may, however, extend some distance further down in places where the tide can no longer have free play. This occurred in the Seine estuary (fig. 68). The authorities of Havre, which lies at one side of the estuary not far from its mouth, feared that if the training walls were brought further down, the deposits might extend to their neighbourhood. The reduction in the capacity of the estuary, due to the deposits, caused it to become filled up more quickly, and the time of high water at Havre was advanced. The dotted lines show a good arrangement of training walls proposed by Harcourt.

There is no doubt that it is always feasible to carry training walls right through an estuary, or at least down to a point where deep water is reached, and if a proper funnel shape is given to the channel the reduction of the tidal flow and silting up of the spaces behind the walls need not cause any trouble. Training the complete estuary was carried out in the case of the Tees, where, however, the estuary was not of great length, and was not of a good shape for keeping itself open. Any affluents entering the estuary can be provided with separate trained channels. Difficulty may, however, arise if there are towns which would be shut off from the estuary by the silt banks.

Generally the line selected for the trained or dredged channel should, though it must be as short and direct as possible, coincide as nearly as possible with that which the water naturally tends to keep open. This may be toward one side of the estuary or the other, according to the direction from which the tidal wave approaches. In the case of the Dee, the best line was not adopted, attention having been chiefly given to the question of silting up the spaces outside the walls and so reclaiming land, a matter which should always be treated as of quite secondary importance. Training walls in estuaries are generally built only up to half-tide level. Were it not for the expense they might be built up to high-water level. In the Seine estuary the walls were made of blocks of chalk.

Whether a trained channel will keep itself open or will need periodical dredging depends, of course, on the amount of silt in the water and on its velocity and depth. The question must be worked out and calculated as in the case of a non-tidal river.

The estuary of the Mersey differs from most others. Towards the mouth, near Liverpool, it is narrow and it widens out further inland. The tides, running through the narrow portion, to fill up the large inland basin and to empty it again, keep the narrow part scoured to a great depth. It was proposed to train the wide portion for the Manchester Ship Canal. The training would, no doubt, have succeeded, but, owing to the silting up of the greater part of the estuary, the scouring near Liverpool would have been very greatly reduced and serious damage done to that port.

CHAPTER XV

RIVER BARS

1. **Deltaic Rivers.**—When a river flows into a tideless sea its silt deposits and forms a shoal or bar. This shoal may in time extend and rise up to the water-level. The current of the river makes its way through it in various directions, and in this way a delta is formed and constantly extends seawards. This flattens the slope of the lower portions of the river, and causes raising of the bed in the reaches upstream, and this again may cause the water to break out further upstream and form fresh channels to the sea. The bars at the mouths of deltaic rivers are generally formed with great rapidity, and they are apt to form a complete hindrance to navigation. They are sometimes partly scoured away by floods in the river, but in this case the scoured material may deposit on the outer slopes of the bar. If a river which carries silt has no delta, it is probably because there is a littoral current, which prevents the silt from depositing. On the other hand, if a river brings down very heavy sediment, a delta may be formed even when tidal flow is not wholly absent. This occurs in the case of the Ganges.

The bars at the mouths of deltaic rivers cannot usually be kept down by dredging except at great expense.

The usual method of dealing with them is to run out two parallel jetties, in continuation of the river banks, so as to bring the mouth of the river out to the bar. The river then scours a channel through the bar and, if the walls are not too far apart, the depth will probably become as great as in the river and sufficient for navigation. The river, however, tends to at once form a new bar further out. The rapidity with which the new bar forms will be greater or less as the specific gravity of the materials carried by the river is greater or less, and as the strength of any littoral current is less or greater. Clay is spread far out while sand quickly sinks. All deposits are, however, swept away if there is a strong littoral current. The steeper the slope of the bed of the sea away from the bar the longer the new deposit will take in forming a fresh bar. Also the less the discharge of the river the less the deposit will be. The branch of a deltaic river selected for improvement by having the bar at its mouth removed, should be one which has a small discharge and whose mouth is in a position where there is a strong littoral current. In the case of the Rhone, the branch selected was the eastern one, whose mouth was not exposed to any littoral current. Moreover, the other branches of the river were closed, and this increased the discharge of the branch which was left open. The work did not succeed. In other cases, the parallel jetty method has succeeded, and notably in the case of the Mississippi. In this case willow mattresses weighted with stones were used. The question of keeping down the discharge does not, however, appear to have always received sufficient attention. In the case of the Mississippi the "South pass" was selected for improvement. In order to remove a shoal its upper end was narrowed and its discharge reduced. The upper ends of the other "passes" were then obstructed so as to restore the discharges of all the passes to their former amounts. The wisdom of this step is questionable. It is desirable to keep down the discharge of the branch which is to be improved to the lowest limit consistent with free navigation.

If the width of the river near its mouth is greater than is desirable for the width between the jetties, the latter are sometimes made to converge though their outer ends are made parallel.

In the case of the Mississippi the jetties were made with a slight curve to the right. It would seem desirable always to make the jetties with quite a considerable curve. The jetty which was convex to the channel could then probably be shortened. In a case where there is a littoral current, say to the right, the curve of the jetties could be to the right, so that the stream on issuing would tend to merge into the current and assist it.

2. **Other Rivers.**—It often happens that the materials —sand, gravel, and shingle—of which a sea beach is composed shift gradually along the shore. This is known as "littoral drift." It is by some supposed to be due to the action of the tides, and by others to the action of waves, the drift taking place in the direction of the prevailing winds, excluding those which are off shore. The latter cause is the more probable.

Most rivers have bars at their mouths. In the case of deltaic rivers the bar, as already stated, is caused by the heavy silt carried by the river, though it may be assisted by littoral drift. In the case of non-deltaic rivers flowing into tideless seas, the quantity of silt is not enough to form a bar, and the same is generally true in the case of tidal rivers where the volume of tidal water is usually much greater than that of the upland water. In both these classes of rivers the formation of the bars is due chiefly to littoral drift or to sediment brought in by the sea water. The bar, as in the case of deltaic rivers, may be partly scoured away by a flood in the river, and the scoured material may deposit on the seaward slope of the bar. Generally, the navigation channel across a bar of this kind can be kept sufficiently deep by dredging, but sometimes jetties, like those mentioned in the preceding article, have been constructed, and in this case there is the great advantage that the bar is not liable to form further out. If littoral drift tends to accumulate, the jetties, or at least the one on the side whence the drift comes, can be lengthened. This was done, as mentioned by Harcourt (*Rivers and Canals*, Chap. IX.), in the case of the rivers Chicago, Buffalo, and Oswego, which flow into the Great Lakes of America. The same writer states that the jetties at the Swine mouth of the tideless river Oder were made to curve to the left, the convex or left-hand jetty being the shorter, but that this exposed the mouth to littoral drift coming from the left. The river, upstream of the jetties, had a slight curve towards the left, but this could have been corrected or, at all events, the jetties made to curve to the right.

A case (fig. 69) where parallel jetties were recently constructed in a tidal sea is that of the mouth of the Richmond River, New South Wales (*Min. Proc. Inst. C.E.*, vol. clx.).

In the case of a bar at the mouth of an estuary, parallel jetties would be too far apart. In such cases converging breakwaters (fig. 70) are sometimes made, especially if the tidal capacity of the estuary is small. The entrance is generally 1000 to 2500 feet wide. If

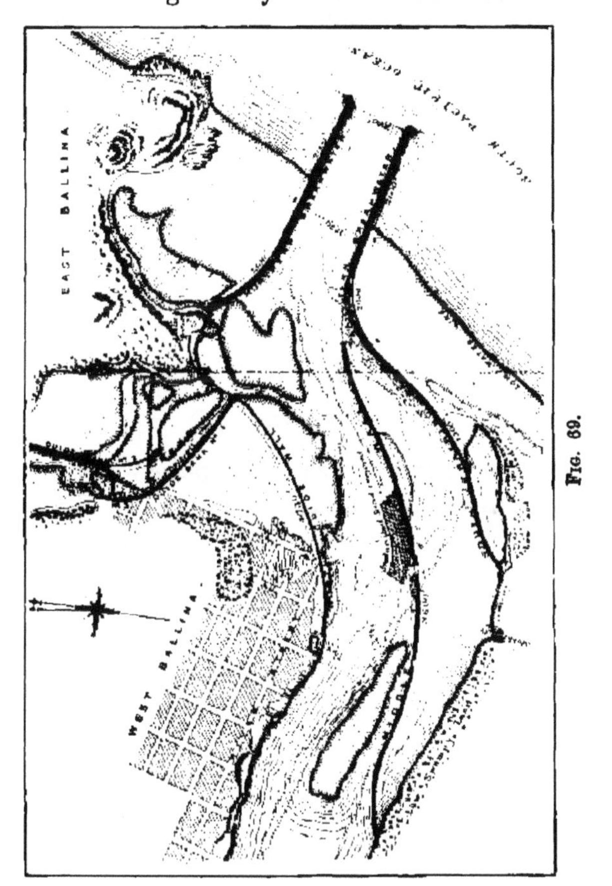

Fig. 69.

made narrow, it would reduce the tidal flow too much. The space inside the breakwaters adds to the tidal capacity, and thus induces scour at the bar. The case is similar to that of the Mersey estuary (Chap. XIV., *Art. 5*), the breakwaters assisting scour at the bar, though perhaps slightly interfering with the tidal flow in the estuary.

Converging breakwaters also tend to stop littoral drift, and the space inside them acts as a harbour of

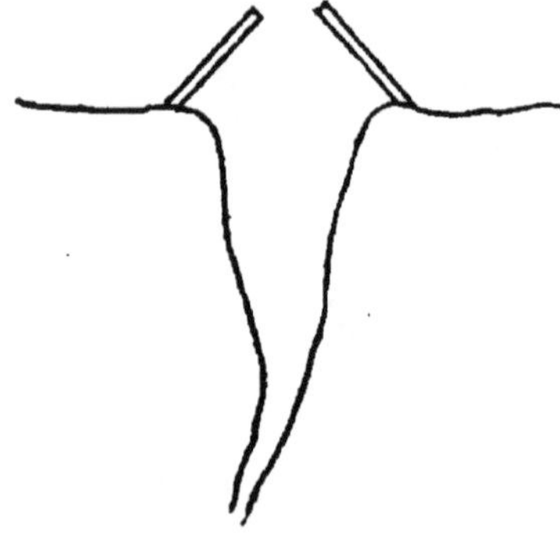

FIG. 70.

refuge in storms and as a sheltered place where dredgers can work (*Rivers and Canals*, CHAP. XI.). They have to be heavily built and are very expensive, and they are generally adopted only when there is an important seaport, and when they can be put to all the uses above indicated.

APPENDIX A

Fallacies in the Hydraulics of Streams (Chap. I., *Art. 4*, and Chap. VI., *Art. 2*).—In an inundation canal in India the supply during floods was excessive. Orders were given that a flume be made at the head, as shown in fig. 71. The sides were to be revetted, as shown in fig. 19 (Chap. VI., *Art. 3*); the length, excluding the splayed parts, was to be 200 feet, and the floor

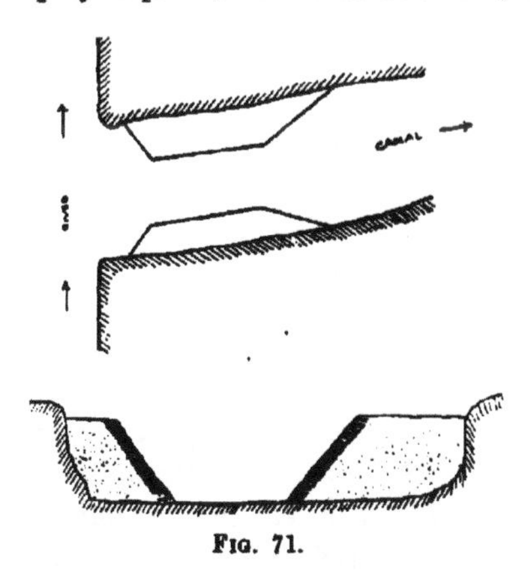

Fig. 71.

was to be a mattress well staked or pegged down. The order stated that "by this means we cannot get into the canal much more than its true capacity." With 9 feet of water, a surface fall of 4 inches in 300 feet would give a velocity of some 6·5 feet per second, and a further fall of about 8 inches would be required at the head of the flume to impress this velocity on the water. The flume would reduce the depth of water in the canal by 1 foot, *i.e.* from 9 feet to 8 feet. This would not be in anything like the proportion desired. Moreover the flume, unless the bed was extremely well protected, would be destroyed. The above is a case of exaggerating the effect of an "obstruction."

Again, on a branch canal it was observed that "wherever cattle crossings exist there is a deep silt deposit which practically blocks the branch." The deposit exists because the sides of the channel are worn down. A wide place always tends to shoal (CHAP. IV., *Art. 9*). If the deposit obstructed the flow of water there would be a rush of water past it, and it could not exist.

The Gagera branch of the Lower Chenab Canal—the left-hand branch in fig. 72—was found to silt. It was pro-

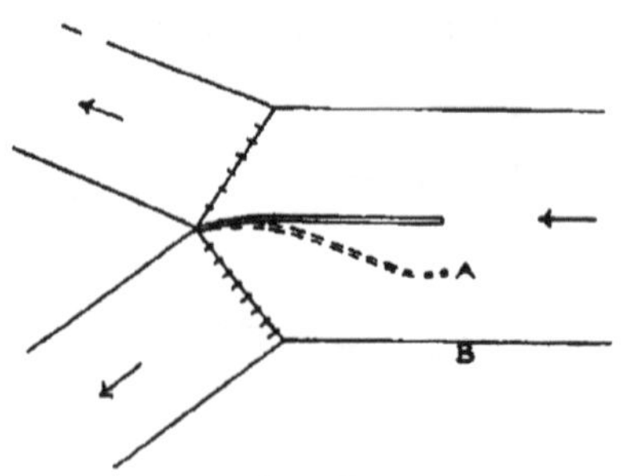

FIG. 72.

posed to make a divide wall (fig. 72) extending up to full supply level. The idea is unintelligible. The silt does not travel by itself but is carried or rolled by the water. As long as water entered the Gagera branch, silt would go with it. The authorities, who had apparently accepted the proposal, altered the estimate when they received it, and ordered the wall to be made as shown dotted and of only half the height. This was done. The idea seems to have been that the wall would act as a sill and stop rolling silt. This is intelligible, but see CHAP. IV., *Art. 2*, last paragraph. Moreover, there was a large gap, A B, in the wall. The work is said to have proved useless, and proposals have been made to continue the wall from A to B. In this form it is conceivable that it may be of use.

In a river, the rises and falls at different places are not, of course, the same, even when they are long continued. In the river Chenab, at the railway bridge at Shershah, the rise from low water to high flood is generally a foot or two more than the rise at a point 25 miles upstream. It has been suggested that the railway embankments, which run across the flooded area, cause a heading up of the stream. If this were the case, to any appreciable extent, there would be a "rapid" through the bridge, which, if it did not destroy the bridge, would at least be visible and audible.

The exaggerated ideas which often prevail regarding the tendency of a river, when in flood, to scour out a new channel, have been mentioned in CHAP. IV., *Art. 8*. Spring, in his paper on river control, admits, when mentioning Dera Ghazi Khan, that there was little danger, but in mentioning the Chenab Bridge at Shershah he quotes, without disputing it, an opinion of the opposite kind (*Government of India Technical Paper*, No. 153, "River Training and Control on the Guide Bank System").

For some other fallacies, see *Hydraulics*, CHAP. VII., *Arts. 9* and *15*.

APPENDIX B

Pitching and Bed Protection (CHAP. VI., *Art. 3*, and CHAP. X., *Art. 2*).—Any scour upstream of a weir is merely due to the eddies formed upstream of the crest (*Hydraulics*, CHAP. II., *Art. 7*), and is not serious. And, similarly, as to scour upstream of a pier. A hole formed alongside a pier or obstruction, if there is no floor, may work upstream. The chief use of a floor extending far upstream is to flatten the hydraulic gradient (CHAP. X., *Art. 3*).

For pitching of the sides, monolithic concrete is not very suitable, because it may settle unequally and crack. For heavy pitching, concrete blocks can be used. They can rest on a layer of 3 to 6 inches of rammed ballast or gravel. The toe wall, as shown in fig. 13, page 65, is sometimes dispensed with, the pitching being merely continued to a suitable depth below the bed, and the bottom edge being at right angles to the slope instead of horizontal. The portion below the bed may be of concrete.

INDEX

PRINTED BY NEILL AND CO., LTD., EDINBURGH.

Lightning Source UK Ltd.
Milton Keynes UK
UKHW022018310822
408147UK00003B/172